Anonymous

Life and Adventures of Sam Bass

Anonymous

Life and Adventures of Sam Bass

ISBN/EAN: 9783744694681

Printed in Europe, USA, Canada, Australia, Japan

Cover: Foto ©berggeist007 / pixelio.de

More available books at **www.hansebooks.com**

Life and Adventures

of

SAM BASS

The Notorious Union Pacific and Texas

Train Robber

Together With

A Graphic Account of His Capture and Death—

Sketch of the Members of His Band, With

Thrilling Pen Pictures of Their Many Bold and

Desperate Deeds, and the Capture and Death

of Collins, Berry, Barnes, and Arkansas Johnson.

Dallas, Texas
Dallas Commercial Steam Print,
1878

CHAPTER I

EARLY LIFE OF BASS.

Hoosier Boy.—Out in the World.—Down the Mississippi.—Rapid
Progress in Dissipation.—Arrival in the Lone Star State.

Sam Bass was born July 21st, 1851, in Lawrence county, Indiana, on a farm two miles north of Mitchell. He was the son of Daniel Bass, who in 1840 had married Elizabeth J. Sheeks and settled upon a farm, where by industry and economy he acquired a competency. They had ten children, the two oldest of whom died in infancy. The third son, Geo. W. Bass, enlisted in the 16th regiment Indiana volunteers and was killed in the battle of Richmond, Ky., August 30th, 1862. The rest of the children, with the exception of the wayward subject of this sketch, are still living in Lawrence and Martin counties, Indiana. Their names are John, Denton and Sallie Bass, Euphema Beasley, Mary Hersey and Clarissa Hersey, "all doing well," so the chronicles state, "and highly respectable."

In 1861 Bass lost his mother, but a few months after his father made haste to supply the loss by marrying "a pious young widow and devoted member of the M. E. Church." This union seems to have been attended by excellent results, as we are informed that soon after the marriage "Daniel Bass joined the M. E. Church and was a praying Methodist up to the time of his death which occurred February 20th, 1864." One child was born from this last marriage, Charles Bass, who now lives at Kansas City, Missouri.

After the death of the elder Bass, Daniel L. Sheeks, an uncle of the heirs, took charge of the estate and the children. Mr. Sheeks, being one of the largest, and most respectable farmers in the country, trained the children up to the habits of industry and gave them all the advantages of education and improvement conferred upon his own children. But as Bass could not read and could barely sign his name, these advantages could not have been great.

There seems to be no question that the Bass and Sheeks families were highly respectable and had the esteem of all the people with whom they lived.

Up to the time of the death of his father and for two or three year afterward, young Bass maintained an excellent character, but after that he began to associate with bad companions and soon acquired evil habits.

In 1869, tired of the restraints of his guardian or longing to see more of the world, he left his Hoosier home and went to St. Louis, a very bad place to go to at any time, as everybody in Chicago would testify, even upon oath. But the great city had but few attractions for the country boy, and hence he took passage on a steam-boat and floated down the Mississippi, landing at Rosedale, Mississippi. Here the young adventurer remained a year, working at Charles' Mill and forming bad habits with reckless rapidity. It is said that he became an expert at card playing and revolver shooting and was noted for his dissipation.

In 1870 he bundled together his little effects again and left for the Lone Star State, arriving at Denton in the latter part of the year. His advent in the Empire State seems to have had a subduing effect upon his mind, for he at once sought employment and began a sober, industrious life. His first engagement was with Mrs. Lacy, proprietress of the Lacy House in Denton. He continued in her employ about a year and a half, giving entire satisfction and greatly endearing himself to the good lady of the house by his kind and obliging disposition and his excellent conduct.

He was next employed by a man named Wilkes, and shortly after this by Sheriff W. F. Eagan, with whom he remained until the beginning of his downward career. Sheriff Eagan speaks of him as a very sober and industrious young man. He frequently entrusted him with considerable sums of money to go to Dallas and other neighboring places to purchase lumber and supplies. His habits of economy were so great that his employer found fault with h'm for starving himself and team. He never would wear a suit of clothes that cost more than five dollars. In all his service he was very much devoted to his employer's interests. He was also retired and quiet in his disposition, never was absent from home in the evening or away on the Sabbath unless sent upon an errand. His only companion was a little boy, who taught him to write and assisted him in his efforts to make a man of himself.

But unfortunately for himself, on an evil day in 1874 he became the owner of a "little sorrel mare." This was the beginning of a downward career, which has made Bass one of the most noted criminals of this or any age; for the mare proved to be fast and Bass soon became faster than the mare. After he had run a few races around Denton his employer saw that driving a team and running races would not go together. Hence he told Bass that he must take his choice between the race mare and the team. Bass at once concluded to keep the mare and abandoned the employment in which he had been industriously and respectively engaged for four years. This was in 1875, and from that time on he gave himself up to a life of dissipation. Soon afterwards he made the acquaintance of Henry Underwood, who became his boon companion, and later was one of the most noted of his gang.

His evenings were spent at saloons, all business was neglected and he was constantly with wild, reckless fellows. His neighbors say that he became wholly unlike himself. This remark explains much of his career; for he evidently had one of those head-strong earnest natures which do nothing by halves. Whatever he did was done with all his might. As a boy at cards he became the most skillful of all his companions; as an employe he was faithful to his employer; as a bandit he outstripped all the daring characters who have wrought deeds of violence upon Texas soil.

It is said that when he was but nine years of age he witnessed a noted criminal trial at his home in Indiana. It apparently made a great impression on his mind and may have excited an evil passion for notoriety, even if it was infamous; at all events he is said to have remarked one day in Denton, when seeing some horse thieves sent to the penitentiary, that when "he committed a crime it would amount to something. He would never be sent to the penitentiary for so small a thing as stealing a horse."

CHAPTER II

FATAL MISTAKE

Working on a Farm—Horse Racing—Beating the Indians—
Texas Cowboy.

It was in March, 1875, that Bass left the employ of Sheriff Eagan. After horse racing, gambling and dissipating for some time in and around Denton, he went to Fort Sill, accompanied by five or six companions. He was absent on this trip two or three months, but what transpired during the time is not known, though it is not believed that he carried his dissipation beyond horse racing and other forms of gambling.

When next heard of he was in the Indian Territory, that beautiful paradise of nature, the government's home for civilized Indians and the hiding place of uncivilized white men.

It is not to be supposed that these races, like the great Ten Broeck-McCarty fiasco, drew all the governors, senators, members of Congress, and other people in the whole region round about, nor is there any record of a heavy gate fee for the privilege of a grand sell. But in spite of his rudeness the red-skinned racer is up to the tricks of the profession and has no trouble with a superabundance of honest scruples.

Bass found it easy enough to beat their scrubby little courers with his sorrel mare. But how to get possession of the ponies he had won was a much more difficult matter. It was in vain that he reasoned, cursed and threatened. They were not playing a losing game, they had the ponies and meant to keep them. But Bass was equally determined to have what the mare won. Therefore when night came he took all the ponies he had won and as many more as he could get his hands on and started for Texas. This is the first act of robbery recorded in his career. The ponies were driven across the State toward San Antonio, where Bass arrived in the latter part of '75, or early in '76.

Here horse racing and gambling were resumed, the sorrel mare still doing the honors of the course. But shortly afterwards this fatal piece of horse flesh, which had so rapidly carried her owner down the course of ruin, was sold. Bass remained in and about San Antonio during the summer, but nothing of note occurred until about August 1st, when he joined Joel Collins, afterwards leader of the gang which robbed the Union Pacific train, in gathering up a drove of cattle for the Northern market. The nucleus, at least, of the drove was purchased, but how many mavericks, old or young, with or without brands, voluntarily or otherwise, slipped into the drove as it moved across the country,

no one can tell. But very loose notions on such points prevail in
the stock ranges, and it is not to be supposed that Collins and
Bass played a puritanical part as they followed their herd across
the wide prairies.

The cowboy is a sui generis of the Southwest. Usually he
is tall and slim, with sunburnt face, keen glittering eye and handsome moustache. His dress is of the simplest kind. A half acre of
hat, more or less, covers his head. His feet are enclosed in a
heavy pair of cow-hide boots, at the heels of which are a pair of
clanking, clattering spurs, and in the tops he stows away the sur-
plus ends of his pantaloons. His shirt is a flannel or calico and
abhors mansard collars, and other "neck fixins." He rides a tough
little animal called a Texas or Mexican pony, which he purchases
for fifteen or twenty-five dollars. It can stand more riding and
less feeding than almost any kind of horse living. At his saddle is
strapped a Sharp's or Wincester rifle, and at either hip is a six
shooter, while around his waist is a belt filled with cartridges. In-
tervening space is filled up with daggers, bottles or whiskey and
plugs of tobacco.

When on the drive he stops wherever night overtakes him,
sleeping as sweetly with his body stretched upon the greensward,
his head upon his saddle and the glittering stars above, as if tuck-
ed away in the softest bed. He always stops at "the store," takes
a drink, buys some more tobacco and replenishes his bottle.

When he arrives in a town or city, if he is flush, he always
finds his way to a gambling saloon, where he plays more recklessly
than successfully, or to a house of prostitution, where he falls an
easy victim to the blandishments of some fair enchantress, and usually retires heavy in head and light in pocket.

But with all his weakness, he has some of the best qualities
of manhood. He is generous, brave, and faithful to his friends, sel-
fishness and small meanness find but little place in his everyday life.

There is no doubt that, as a cowboy, Bass snugly filled out
the proportions of the type.

But not to make a further disgression, we find that the drovers
arrived in Kansas some time during the fall. Here they sent their
cattle on to Sidney, Nebraska, while they took the cars for the
same place. It is said by detectives that his was done because
the ownership of some of the cattle was a disputed question, and
that they were afraid at that time and place to be seen with them.
But it is quite probable that they were tired of the drive and took
this method of obtaining a rest.

At Sidney they met the herd and drove it from there to the
Black Hills, where it was disposed of.

This ended the cattle business and introduces us to another
phase of the bandit's career.

CHAPTER III

LIFE IN THE BLACK HILLS

On the Road—Keeping a Dance House in Deadwood—Deadwood
Society—Belles of the Town—Kitty Leroy, the Danseuse—Her
Tragical Death—"The World by the Tail, With a Down Hill Pull"
—The "Tail Holt" Slips

After disposing of the herd of cattle, Bass and Collins purchased
two four-horse teams, and began freighting between Dodge City,
Yankton and the Black Hills. In this business they continued until
January, 1877, when they sold out and opened a gambling saloon and
house of prostitution in Deadwood.

There is nothing puritanical or bigoted about Deadwood society.
The widest latitude of opinion and practice is allowed on all—moral
questions. The conscience is not harassed with scruples and no pru-
dential considerations harness the passions. Nobody seems to have
the slightest recollection of a father's solemn admonitions or a mother's prayers. Religious teaching is a withered tradition, tossed among
the other rubbish of abandoned sobriety. Sunday is no better than
any other day, and every other day is as bad as it can be, but night
is still worse.

Every man who goes to Deadwood is shadowed by the presentiment
that he will either be shot or that the mad fever in his blood will
break out in the slaughter of somebody else. When he arrives in
the city he needs no introduction, but to hang out his revolvers, call
for a drink and lay down a greasy pack of cards. He is asked his
name, for convenience sake, but nobody thinks of inquiring where
he came from, why he left or what his name was before he left.

Most of the houses are saloons. The rest are theatres, faro
banks and dance houses. Prostitution is not confined to special
quarters but has full sweep of the range. Only respectability and
virtue are crowded into corners.

The queens of society are the most brilliant of the demimonde.
The further they have fled from the modesty of their sex, the more
dashing and daring they are, the more recklessly they can handle a
revolver and the straighter they can throw a dagger; the more men
rave over them and the more ready they are to kill or be killed for
their sake.

To show that this picture of Deadwood society is not overdrawn,
and to present a fair type of the leaders of the sex, we give below a
description of one of the queens who reigned in the height of her

glory at the time Bass and Collins kept the dance house. This was
Kitty Leroy, a woman who has been much written about and whose
tragical fate shortly after this sent her name throughout the press
of the country. A Black Hills letter speaks of her thus:

Kitty LeRoy, who was killed by her husband only a short time ago,
who then killed himself, was a small figure, and had previously been
noted as a jig-dancer. She had a large Roman nose, cold, grey eyes,
a low, cunning forehead, and was inordinately fond of money. I
saw her often in her "Mint," which was opposite my office, where
men congregated to squander their money; and as Kitty was a good
player, like the old grave-digger, "she gathered them in!" that is,
their money. Men are, in a general sense, fools. A small tress of
golden hair, or a bright eye or soft cheek will precipitate them into
an ocean of folly, and women of the world (and some out of the
world) know this fact and play upon the weak string of men's hearts
until all is gone—money, character and even life. Kitty had seen
much of human nature, entering upon her wild career at the age of
ten. She was married three times and died at the age of twenty-
eight. A polite and intelligent German met her. He was going well
with his gold claim; she knew it. Like the spider, she spun her
delicate web about him until he poured into her lap $8,000 in gold,
and then when his claim would yield no more she beat him over the
head with a bottle and drove him from her door. One and another
she married, and then when their money was gone, discarded them
in rapid succession. Yet there was something peculiarly magnetic
about Kitty. Men did love her and there are men living to-day who
love her memory. Well, she's gone. I saw her only a short time
since, lying dead by the body of her inanimate husband, with whom
she would not live, but with whom she was obliged to pass quietly
to the grave.

Another correspondent writes of Deadwood society and very
gushingly of Kitty as follows:

"There are dance houses and theatres, where the gay society
congregates, and it is at such houses, as well as at the gambling
houses, that the fair sex may be seen. The women, though not so
bad as the men, are all strong minded, which, from a hen-pecked
point of view, is the worst thing you can say of a female. Some keep
bars, taverns, boarding houses, and variety shows, while a few keep
gambling dens, like 'The Mint,' which was kept by poor Kitty LeRoy,
lately killed by one of her husbands, which was the tragic end of a
brilliant career; for, barring the wild, Gipsy-like attire, which fash-
ion would fail to appreciate as intensely picturesque, Kitty LeRoy
was what a real man would call a starry beauty. Her brow was low
and her brown hair thick and curling; she had five husbands, seven

revolvers, a dozen bowie-knives and always went armed to the teeth, which latter were like pearls set in coral. She was a terrific gambler, and wore in her ears immense diamonds, which shone most like her own glorious eyes. The magnetism about her marvellous beauty was such as to drive her lovers crazy; more men had been killed about her than all the other women in the Hills combined, and it was only a question whether her lover or herself had killed the most.

"She could throw a bowie-knife straighter than any pistol bullet except her own, and married her first husband because he was the only man of all her lovers who had the nerve to let her shoot an apple off his head as she rode by him at full speed. On one occasion she disguised herself in male attire to fight a man who had declined to combat with a women He fell, and she then cried over him, and married him in time to be his widow. Kitty was sometimes rich and sometimes poor, but always lavish as a prince when she had money. She dealt 'vautoom' and 'faro,' and played all games and cards with a dexterity that amounted to genius."

Kitty is supposed to have been the wife of Capt. E. H. Lewis, of Bay City, Michigan. But in 1872 she left her husband, and after that time figured as a public dancer in various part of the Union. In the winter of 1875-6 she was engaged at Thompson's variety den in Dallas. While there she created quite a sensation among the lewd habitues of that resort, by her artistic dancing and gay rollicking and dashing manners. After a few months stay she ran away with a well known saloon man, and together the two visited California, where they remained a few months and then proceeded to Deadwood. Subsequently she quarreled with her paramour and married Samuel R. Curley, a note faro dealer. But the couple proved to be badly mated, and soon after their marirage Curley went to Denver, and almost immediately thereafter the broken friendship between Kitty and her paramour was restored, a fact that was communicated to Curley, who undoubtedly went to Deadwood for the express purpose of killing his wife, her paramour and himself, for he traveled under an assumed name; alighted from the coach in South Deadwood, telling the driver if asked if any passengers other than those delivered at the office had come up, to say no. He walked direct to the hotel at which the unfortunate woman was a guest, remained there all day, and in the evening sent for his rival, who refused to to visit him. He then told a colored man employed in the house that he intended to kill his wife and himself, and true to his word went up stairs and did so.

When the bodies of the murderer and his victim were found the woman rested upon her back, in a position and with a quiet facial expression that indicated naught of the bloody deed that had been enacted but a moment before. Close examination revealed a small bullet hole in the waist of her dress, which, upon being opened, disclosed the fatal wound in the center of her chest. In the opposite corner of the room lay the murderer upon his face, in a sickening pool of blood, his brain oozing out and pieces of skull protruding from a ghastly wound. His right arm was doubled up behind him, the hand grasping a Smith & Wesson, by which the fatal deed was committed.

It is not strange that in such society as this Bass soon became fit for crimes of the first magnitude. But before entering on his career of daring deeds he seems to have made one more effort to follow a respectable occupation. For about this time he wrote to Henry Underwood, then in Denton, that he and Collins had purchased a quartz mine, for which they had been offered $4,000, but it was a big thing and he would not sell, but when he got it worked up he would let him know all about it. He assumed his friend in the phraseology of the cattle ranche, that he had the world by the tail, with a down-hill pull. He also informed him that he would return to Texas in the Fall and would then pay off his creditors. But whether the "tail holt," of which he boasted, slipped or whether the mine had been salted for his special benefit by men shrewder than cow boys, is not known. But this was the last respectable piece of business in which he engaged. After this he is known to the world only as a bold highwayman, undertaking deeds of daring which but few bandits have the audacity to attempt or the nerve to execute.

What led immediately to his final plunge into a career of bold outlawry is known only from the statement which he made on his death bed. When asked the question why he began to rob he replied that he had won some money gambling and had been robbed of it and wanted to get even with them. Whether this occurred at Deadwood or elsewhere is open to conjecture, but there is some reason to believe that he referred to some occasion on which he had been beaten out of his money by sharpers. The fact that both he and Collins got rid of their cattle money so fast, indicates that if there were none more reckless and daring, yet there were much shrewder gamblers and sharpers in the Hills than Bass and Collins.

CHAPTER IV

LAUNCHING INTO CRIME.

Black Hills Stage Robberies—Great Union Pacific Robbery At Big Spring, Nebraska—Sixty Thousand Dollars In Gold Captured.

While in the Black Hills Bass made the acquaintance of Nixon and Jack Davis, and probably of all the men who assisted in the Union Pacific train robbery. But according to his own statement only Jack Davis and Nixon were engaged with him in stage robbing.

There were seven of these robberies in all, and some money was realized from them, but how much is not known.

It is apparent that but a short time elapsed between the stage robberies and the capture of the railroad train at Big Spring, Nebraska.

This was one of the most daring and successful train robberies ever committed.

Collins formed the plan of the robbery, though it is believed Jack Davis first suggested it. He had come from San Francisco, and was familiar with the fact that large sums of gold were constantly passing over the route. The names of the gang, six in number, have all been ascertained since, Bass himself testifying to the correctness of the list in his dying moments. They were Joel Collins, formerly from Dallas county, Texas; Sam Bass, from Denton county, Texas; Jack Davis, from San Francisco; Bill Heffridge, who went from San Antonio with Collins and Bass; James Berry, from Mexico, Mo., and Nixon, of whose previous history but little is known.

A more daring and desperate band of outlaws was never gotten together in this country. Collins acted as leader of the band and it has been charged that he spent three or four weeks previous to the robbery at Ogalalla, Neb., gambling and associating with desperate men, from whom he organized the gang. It has also been said that he had a cattle ranche near Big Spring Station, and thus became acqainted with the habits of the station men, the operations of trains and the surroundings of the office. But this has been denied.

The time selected for the robbery was Tuesday night, the 19th of September last. As Big Spring was only a water station, the plan evidently was, to capture the few men employed about the station and keep them under guard until after the train was robbed.

At the appointed hour the bandits boldly rode down towards the

station, hitched their horses conveniently near, and at once proceeded to business. With a small flourish of revolvers and the well
known command, "throw up your arms," the station agent and as-
sistant were soon made secure.

As train time, 10 o'clock, drew near, the bright rays of the head-
light were seen falling upon the distant track. Then came the long
sound of the whistle, the rushing train checked its speed and in a
moment more stood still upon the track. It was but the work of
an instant for one of the gang to mount the engine, command the
engineer and fireman to throw up their hands and there hold them
helpless under the muzzle of a cocked revolver. But even before
this had been accomplished, two of his confederates had boarded
the express car and were ransacking its contents. They soon found
a large quantity of gold in one of the safes, but the other could not
be opened. It was in vain that they ordered the messenger to open
it. He assured then that he had no key, that it was a time lock and
could only be locked or opened at each end of the route. Jack Davis
cursed and raved, beat him over the head, thrust his revolver into
his mouth knocking out one of his teeth and lacerating the flesh, and
threatened to blow the top of his head right off if he didn't open it.
But Bass said he reckoned the messenger was telling the truth, and
that they had better give it up.

After going through the coaches and robbing the terrified passengers the bandits slowly backed away, keeping their arms presented
until they were lost to view in the darkness.

A number of shots were fired during the transaction and a few
wounds were inflicted, but no one was killed.

The gold taken from the express car amounted to the sum of
$60,000, no small weight for the robbers to handle under the circum-
stances. It consisted entirely of twenty dollar gold pieces, of the
coinage of 1877, a fact which was afterwards of material assistance
in ferreting out the perpetrators of the crime.

Shortly after the robbery, the gang divided the money and
separated, going two together by different routes. Bass, Davis, and
Nixon, for a time vanished from view. Of the others we shall speak
in the succeeding chapters.

CHAPTER V

FATED.

Great Excitement Among Railroad Officals.—On The Trail —Uncle Sam's Soldiers to the Front.—Two Travel Stained Cow Boys.— Heavily Laden Pony.—Capture of Collins and Heffridge.—They Die With Their Boots On.—Subsequent Doubt.—Collins Defended.—Short Sketch of His Life.

The Big Spring robbery created intense excitement among railroad officials and caused a general sensation throughout the country.

The large amount of money secured was looked upon as a temptation which would soon lead to another like attempt. This in connection with the heavy shipments of gold over the line would be likely to excite the cupidity of every bandit in the West, at the same time the long stretch of road through waste and desert regions, with here and there a lonely station, made it very difficult to afford adequate protection. It was determined, therefore, to capture the robbers at any cost and hazards. Large rewards were at once offered by the State authorities of Nebraska and by the railroad companies. This brought forward detectives from almost every quarter. Telegrams were sent to all officers along railroad lines, to sheriffs and officers in command of U. S. troops.

At first it was not wholly known who the robbers were or whence they came. But it so chanced that among the passengers on the plundered train was a young man named Andy Riley, a resident of Omaha. During the attack upon the train, Riley stood upon the platform and received a wound in the hand from one of the flying bullets, he was also robbed along with the rest of the passengers. He had traveled with Joel Collins on the way to Deadwood and knew him well. He had also seen him and conversed with him only a few days before, while on a visit to Ogalalla. Great was his surprise, therefore, when the robbers came through the train, to find Joel Collins among them. Immediately upon his return to Omaha he notified the officials of the fact and Collins' name and description of his person were accordingly telegraphed in all directions.

It was shortly learned also that after leaving the railroad, the robbers crossed the Platte river, in Nebraska and were next heard

of at Young's ranche on the Republican river in Kansas.* This was
on the 23d, the next Saturday after the robbery. Intelligence ot
this fact having reached Sheriff Bardsley, of Ellis county, Kansas,
he at once started from Hays City, on the Kansas Pacific road, with
a squad of ten cavalrymen and a detective from Denver and made his
headquarters at Buffalo Station, on the Kansas Pacific. This is sixty
miles west of Hays City, in the center of a wild and dreary waste.
Nearby is a large ravine, in which the Sheriff and his posse camped.
While there, about nine o'clock in the morning of the 26th, Joel Collins, the chief of the train robbers, and a single adherent, rode up to
the lonely station.

The following account of their capture and tragical death we take
from a Western daily of the 28th:

When first seen they were riding from the north, coming boldly
over a high ridge of open prairie. They led between them a pony
heavily laden with something which, while it was not bulky, seemed
to tax the strength of the pony to carry it. The men were dusty
and travel stained. They appeared to be and might have been taken
for two Texas "cow boys" out on a hunt for cattle or on their way
to join a herd. Had they rode straight across the track and continued
their journey without stopping, no suspicion would have been aroused;
but they were led instinctively to their death. They rode their jaded
horses to the shady side of the principal building of the station, and
one of the two dismounted, leaving his partner in charge of the
horses and the pack pony. The man left in charge of the horses said
they were Texas cattle men on their way home, and enquiring the way
to Fort Larned. The dismounted man walked up to the station agent
and enquired the way to Thompson's store. The building was pointed
out to him, but as he stood conversing he took out his handkerchief,
which revealed a letter in his pocket upon which was plainly visible
he superscription "Joel Collins." This was the name of the leader
of the Union Pacific train robbers, and the brands upon their horses
assured the station agent that these were the men wanted by the
Sheriff and his soldiers encamped a few hundred yards away. Sheriff
Bardsley was notified at once, and he came up to the station and
examined the horses and made other satisfactory observances. He
conversed with the robber chief for some time, and asked many
questions, which were freely answered. They walked together to the
station and took a drink, and conversed upon various unconsequental
subjects. Collins made no effort to conceal his real name. He had

---

*Detective Leech, of Ogalalla, afterwards claimed that he was in the camp of
the gang on the night they divided the money, and that he knew by sight all
the robbers. He said that he escaped capture at their hands only by hasty flight
on his horse

no suspicion whatever that the telegraph had given his name and description at that little station in the middle of the buffalo plains. Bardsley then left his prey and started back to the camp of the soldiers, who were under the command of Lieut. Allen, and ordered them to saddle up and follow him, and he would bring back the Texans.

In the meantime the two horsemen with their heavily burdened pony had started out on the open plains southward. Sheriff Bardsley and his posse started out in pursuit.

When Collins and his companions saw the Sheriff and his blue coated posse of cavalry appear on their trail, they manifested no excitement. They did not even attempt to run. On the contrary, they rode on leisurely on the Texas trail unil Sheriff Bardsley rode up and halted them. Even then they gave no sign of trepidation or excitement. Collins looked at Bardsley with the coolest effrontery and demanded his business. Said Sheriff Bardsley:

"I have a description of some train robbers which answers well to your appearance. I want you and your partner to return with me to the station. You need fear nothing if you are innocent, and if you are the man I want, then I am $10,000 better off. Please come back to the station, gentlemen."

"You are mistaken in your men, gentlemen," said Collins, laughingly, "but, of course there is no use to object. We will go back and have the mistake explained. We are Texas boys going home—that's all."

Then they turned their tired horses back towards the station. As they returned they exchanged a few brief words which were undistinguishable even by the nearest trooper. They rode a few hundred yards over the level plain towards the solitary station, when suddenly the leader, Joel Collins, broke the silence. Turning to his companion he said:

"Pard, if we are to die, we might as well die game."

Then he drew his revolver. His partner followed his example, but before either could fire, the troops had fired a volley into them and they fell from their horses riddled with bullets. The robbers died instantly and were taken to the station for burial, but were afterwards taken so Ellis station, where an inquest was held upon the bodies.

The body of Collins was identified by a dozen of his old Texas acquaintances but for a long time the body of his accomplice could not be identified. It was at first believed to be that of Sam Bass himself, and was so telegraphed over the country and published in the papers.

Finally Anna Langs appeared in the depot, where the bodies were lying, and stated, under oath, that she recognized the body as being that of William Cotts, formerly of Pottsville, Pa., but more recently of San Antonio, Texas, and that his father resided in Pottsville. He was a light complexioned man, about thirty years of age, light hair and sandy beard, about five feet seven inches high and weighed 135 pounds.

Whether Anna had any real knowledge of the man, or whether she thought so "because she thought so," or whether the true name of Heffridge was Cotts, is difficult to say.   But there is no doubt now that the man who fell with Joel Collins under a shower of bullets, was the member of the gang known as Bill Heffridge.

At the time of his death Collins was described as being dark complexioned, with black hair and beard, about five feet eleven inches high, weighed one hundred and fifty pounds, and was supposed to be twenty-eight years old.   He was also said to have been affable, of pleasing address, intelligent, and very handsome.   Upon his body a small piece of paper was found upon which was written a poetical effusion by a lady and dedicated to Joel Collins.   But the richest discovery was made upon the pony.   When the tired little animal was stripped of the blanket which covered the pack saddle, an old pair of pantaloons was found underneath.   The ends of the legs had been tied together, then they were filled with gold and thrown across the saddle.   When the glittering metal was turned out upon the ground and counted, it was found that the amount was no less than twenty-five thousand dollars.   It was all of the mintage of 1877 and in twenty dollar pieces.

This fact, taken in connection with the other circumstances, furnished the strongest evidence that the lucky Bardsley had struck the right man.   Railroad officials were in high glee and congratulations were exchanged all along the line.

But in a few days serious doubts began to creep into many minds and it was gravely feared that the deadly rifle had struck down an innocent man.   A leading law firm at Topeka, Kansas, was retained to investigate the circumstance attending the bloody tragedy.   It was alleged in Collins' defense that he obtained a large sum of money from the drove of cattle which had been disposed of the pervious year, that he had written to his father that he had obtained twenty-five cents per pound for them and would soon start home with the money.   It was also stated that he had amassed a considerable fortune in the cattle business with his brother near San Antonio and that this precluded all temptation to commit robbery.   In addition to this it was said there was unquestionable evidence as to the time Collins started for home, and of his movements, which tended to show that he could not have been in the vicinity of Big Spring at the time of the robbery.   Collins' conduct at the moment of his death was accounted for on the supposition that he believed himself in the hands of a gang of outlaws who intended to rob and murder him, and that he was determined to sell his life as dearly as possible.

About this time, also, an old Texan came out with a short newspaper article defending Collins from many of the charges which had been made against him and stating it to be the belief of those who had long known young Collins and his parents, that he was not the guilty man.

But a few days later the dying statement of one of his captured confederates forever set at rest all doubt in regard to the matter. Since then the death-bed statement of Bass himself has been added to the proof.

As Joel Collins, so far as known, participated in but one noted crime, few events in his life have been preserved on record.

He was born in Dallas county, Texas where his parents still reside, his father being a farmer of some means and a man who has long enjoyed the respect and sympathy of his neighbors.

In 1868 Young Collins left home and went to the southwest part of the State where he had a brother in the cattle business. From 1868 to 1870 he was in the employ of Allen and Poole, the great cattle men of the coast, and stood well as a young man.

In 1871 he took a herd of one thousand cattle to Kansas for Bennet and Schoate, of North Texas.

In 1872 he took up a large herd for P. T. Adams, Joel receiving one half the profits. In 1873 he did the same thing and on the same favorable terms for James Reed.

In 1874 he bought a drove from Bennet and Akard, partly on time and was induced (if not forced) to ship them to Chicago at a heavy loss. This he did against his will in order to meet the deferred payment, when the cattle were poor and the market down.

This is the statement made by his friends, while others give a different version of the matter.

In 1875 he kept a saloon in San Antonio for a few months. The house is said to have been a disreputable one.

In 1876, as we have already seen, he took his last drove North in company with Bass. In the Spring of 1877 he is said to have opened a provision store at Polato Gulch, thirty-five miles from Deadwood. His friends claim that he remained there until he "started for Texas." But there is much reason to doubt whether it was known to them what he was doing during his last stay in the North. His letters were not always intended to give the exact situation of affairs. It has been charged that he killed several men during his life, but this is probably an exaggeration as there is no authentic account of more than one such act. In 1869 he killed a Mexican in Victoria county, but surrendered himself, was tried and acquitted. As is well known it is very difficult for an American to murder a Mexican. It is a principle with jurists that such acts are always for self defense.

Before closing this chapter we pause for a moment to note the fatal chain of apparently trivial circumstances which so quickly tightened around the unfortunate Collins. Never was the perpetrator of a great crime stricken down by a more unerring blow of retribution at the very moment when escape seemed well night assured. It must be admitted that he showed a singular lack of shrewdness, first in not thoroughly disguishing himself when he boarded the train, and secondly in not giving a wide berth to all telegraph stations. But still, had he rode through Buffalo Station without stopping he would have passed unnoticed; or had he left the dust and sweat upon his face and allowed his handkerchief to remain in his pocket, or had the tell-tale envelope not clung to it, he might soon have been beyond the reach of detectives and soldiers. But the unseen hand of fate had marked him for her own and at that very hour.

His bold attempts to defend himself against a whole troop of soldiers may be called bravery, but it was the extreme of folly. In a country like this, where jails are weak and the law weaker than the jails, where the whole criminal jurisprudence seems to be run for the protection of criminals rather than the public, it would have been better to submit quietly and await a better opportunity. There is reason, however, to believe that this reckless leader of bandits feared Judge Lynch and preferred to "die game."

CHAPTER VI

JAMES BERRY.

Hot on the Track—Berry in Mexico, Missouri—Sells Large Sums of Gold—Scatters Money With a Free Hand—Whole Corps of Detectives in Pursuit—A Disappointment—Better Luck—The Bandit Captured—His Confession and Death—Escape of Nixon.

The capture and death of Collins and Heffridge occured September 26th, but no further clue to the remaining robbers was obtained until about the 8th or 9th of October, when suspicion was aroused in Mexico, Mo., by a large sale of gold which was made there. While at Boonville October 11th, Col. A. B. Garner, General Superintendent of the M. K. & T. railway, received the following telegram:

To Col. A. B. Garner, Boonville, Mo. :

Mexico, Mo., Oct. 11.—James Berry, an old resident of Callaway county, is one of the Union Pacific train robbers. He was at Williamburg Monday night. He is six feet high, weighs about 190 pounds, is forty years old, has a red face, yellowish red hair, mustache and goatee, just recently shaved, round, full face, blue eyes and freckly hands. We will pay $500 for his arrest and ten per cent of the money recovered. He had about $9,000. Think he is making for Texas. Have all crossings closely watched. He has a pacing bay horse and new saddle.

On the same day the Moberly Monitor published the following:

"A man by the name of Jim Berry, of Callaway county, has just returned from the Black Hills to Mexico, Mo. Suspicion has been directed to him of complicity in the Union Pacific roberry, by a financial transaction in which he was engaged immediately on his return. The morning after his arrival in Mexico he visited the banks at the hour of opening and sold gold to the amount of $9,000. Berry remained in Mexico all day Friday and until Sunday evening. He was princely extravagant with his money. Meeting an old mining acquaintance he gave him $250; he delighted a clothier by purchasing a fine suit of clothing without higgling at the price, and bought a $300 bill of groceries, which were sent to his family in Callaway. Saturday evening he took his departure, and Monday morning the bankers received news that the gold he had exchanged, and which they had shipped to St. Louis, had been identified as part of the treasure captured by the Union Pacific robbers. The next day (Tuesday) a corps of detectives from St. Louis and Chicago arrived at Mexico, and with the Audrain county sheriff at their head, started in pursuit of Berry. After a long rough ride the vicinity of his house was reached and the party so disposed as to completely surround it. They now felt sure of their game and the rich reward that awaited his capture. But they were doomed to disappointment. Narrowing the circle and gradually closing in, a rush was finally made for the house. They encountered no opposition where they had calculated upon a fierce resistance, and, upon entering, they found that the bird had flown. A thorough search of the premises revealed no trace of the daring robber, and, though the whole country had been scoured by different parties, his trail had not been struck up to yesterday There is not a particle of doubt that Berry was one of the robbers and his capture is only a question of time."

This prophecy was shortly fulfilled, as the following account of his capture, published in the Mexico Ledger, October 15th, will show:

"We have just interviewed H. Glascock and J. Berry, concerning the arrest of Berry, Sunday morning, and we give you the facts as near as possible below:

"It appears that last Saturday night as our sheriff was eating supper, about half-past six o'clock, he received a message that a man was in town after the suit of clothes Berry had left at Blum's The man's name was Bose Cazy; he lived near Berry's. He told Blum that Berry had told him that he could have the clothes if he would pay the balance of $30 due on them. This was the way he had his "job" fixed up. Glascock ran right down to Kabrich's hall and hid behind the corner and saw Cazy come out; this was half past seven. Glascock followed him to Wallace and McKamy's livery stable. Just as Glascock got near the stable he met J. Carter, and told him to come along. Carter, Glascock and Cazy all got to the stable at the same time. Cazy paid for his horse feed and started to get on his horse. Sheriff Glascock took Cazy by the collar, presented a pistol to his head and told him he would shoot him if he moved. Cazy did not move. Glascock ordered two more horses saddled. They then tied Cazy on his horse. The sheriff and Carter then got on their horses and the calvacade moved off, Glascock leading Cazy's horse. They went down to the branch near Tom Smith's in South Mexico, and as they thought no one would get wind of them there, they stopped. Glascock then went and got John Coons, Bob Steele and a young man named Moore. All got horses and double-barrel shot guns which were loaded with buck shot. They then told Cazy they would have to know where Berry was. He said he had not seen him since he (Berry) had told him he could have the clothes, which was about a week before. The men started out towards Cazy's house, and passed Jeff Jones about 12 o'clock Saturday night. About three o'clock they got to James Armstrong's. Sheriff Glascock told him what they had done, and he wanted Armstrong to go with them and show them where Cazy lived, as he was afraid that Cazy would fool them. Armstrong said he did not know where Cazy lived, and so would not go. We don't know whether Armstrong knew or not. It was then three o'clock Sunday morning. The posse then all got around Cazy, put their guns to his heart and told him if he led them into any trap, or did not take them at once to his house they would shoot him down in a minute. He said he would take them to his house if it would do them any good. When they got within about a half a mile of Cazy's house they took Cazy off, tied him and left Bob Steele to guard him; then Glascock placed two men north of the house and stable, Moore and himself going to the south and west side, and as the open timber was there they thought he might be over in that. They did not alarm Cazy's house at all, it was not quite daylight yet. They all secreted themselves in thickets, as mentioned above, to await results. Glascock told his men: "Boys if you see him halt him; if he shows fight shoot him down; if he runs shoot him in the legs; catch him at all hazards." In about half an hour Glascock heard a horse "nicker" about a half a mile off, as he thought. Moore and Glascock then crept toward the noise, went 300 yards down the branch, came to a fence, saw fresh horse tracks. Glascock got over the fence and got into a thicket; heard the horse snort about fifty yards off in the brush. Glascock took off his hat and crept up twenty yards closer; then he raised up and saw Berry unhitching the horse from a tree. Berry then led his horse aslant toward Glascock, as Berry now says, to lead him to water. Glascock cocked both barrels of his gun, ran out about twenty yards, within about twenty feet of Berry, and demanded him to halt. Berry started to run; Glascock shot, but aimed too high, which caused the charge to go over Berry's head. He shot again and seven buckshot lodged in Berry's left leg below the knee. Berry fell to the ground. When Glascock got to him he was trying to get his pistol out but he could not get it out before Glascock was on him and snatched it away from him. He then asked Glascock to shoot him, that he did not want to live.

"Glascock told him no; that he did not want to kill him, he wanted him to have justice. Just then Moore came up.

"After Moore came up, Glascock called for the rest of the posse when they all gathered around Berry. Glascock then searched him and found in his belt five $500 packages, and in his pocketbook was found $340. He had a gold watch and chain, one dressing coat, three overcoats and comfort. He had doubtless slept there within ten feet of the horse. They took him to Cazy's house, when Mrs. Cazy got breakfast for the men, while a messenger was sent to Williamsburg for medical assistance.

Immediately afer breakfast Sheriff Glascock and John Carter started for Berry's house to look for the balance of the money. Upon arriving there Glascock inquired of Mrs. Berry the whereabouts of Berry; she replied that she did not know, as she had not seen him for four or five days, and thought he had left the country. Glascock then showed her the watch and chain, when one of the children said: Oh, I thought that was papa's. Glascock then told her he had got Berry, when she asked if he had been taken alive, and receiving an affirmative reply, said: I never thought he would be taken alive. He has said a good many times he would never be taken alive. At this they all began to cry—the wife, one little boy and five little girls. It was a very distressing scene.

"Glascock searched the house, but found no money. The house was well provisioned for the winter—hams without number, sacks of flour and coffee, kegs of molasses, etc.

After Glascock left Cazy's about forty of Berry's friends came around and made threats about taking him away, but they did not make any attempt at all; it all ended in talk."

At first it was not thought that Berry's wounds were serious, but gangrene set in and on the night of the 18th his sufferings became very great. It was apparent that he had not long to live, but he maintained a determined and bravado spirit to the last. As the deep silence of night settled down upon all without, he lay in his gloomy cell, alternately writhing in paroxysms of pain or coolly talking to the officers who remained with him to the last, anxious to secure a dying confession. This wish was partly gratified, as he stated in his dying moments, that he was one of the parties who committed the Big Springs robbery; that Collins had planned the robbery and that the names of the rest of the gang were correct, as given by the Express Company. He said that they all traveled together two hundred miles and then separated in squads of two, that his partner came to Mexico with him and then went on to Chicago. This partner must have been Nixon, as it is well known that it was not Davis. After much suffering Berry died at one o'clock in the morning. He left a wife and six children. He is said to have been very respectably connected in Gallaway county.

Thus it is seen that in less than a month three of the Big Spring robbers had been consigned to bloody graves. They had lost their booty and paid the penalty of their crimes with their lives.

There is little reason to doubt Berry's statement that Nixon went to Chicago. From that city he probably went to Canada, as Henry Underwood stated to the officers at Omaha, December 30th, that Nixon was a Canadian and that he was then in Canada. Bass also made the statement at Round Rock.

CHAPTER VII

ESCAPE OF BASS AND DAVIS

Four Days With Their Pursuers—They Separate—Davis Goes To
New Orleans and Bass to Denton—Trip to San Antonio—Underwood Wanted for Nixon—Failure to Capture Him—The Curse
of Stolen Gold.

According to the dying statement of Davis, the whole gang
traveled together two hundred miles and then separated. Where
this separation occurred is not known, though it was probably
somewhere near the Republican river in Kansas, where the gang
was seen four days after the robbery.

In the separation Bass and Davis chose to go together. Like
the ill-fated Collins, they also started for Texas, but by a different
route. It is said that they visited Sidney, Neb., after the robbery,
and left the city suddenly. But this is doubtful, as it would have
been attended by very great danger. Be this as it may, it is certain that they purchased a one-horse hack, loaded their gold in it
and turned their course southward just as soon as circumstances
would permit. Months after this when Bass was safe in Denton,
while lounging the day away in camp, he told the boys, that soon
after they set out in the buggy they fell in with a company of
soldiers and detectives. They at once assured them that they too
were detectives hot in pursuit of the bold bandits, who had robbed
the Union Pacific train, and that they hoped to come up with
them, for there would be a big thing in the capture. This threw
the officers off their guard, and the two were allowed to join the
squad. They continued with them four days, while twenty thousand dollars of ill-gotten gold clanked under their seat as the old
hack rattled over the road. The officers with whom they laughed
and joked all day long, would have given thousands of dollars to
have known this secret, but it remained concealed, and finally the
wily robbers bade them good bye and drove gayly away.

The next heard of them was in Cooke county, where Bass passed
under the name of Samuel Bushon. Here they separated, Bass
going to his old home in Denton and Davis departing for the gay
metropolis of the South. Here he seems to have lived a fast life,
spending his money freely, and enjoying himself as passion led or
vice dictated.

But it was not long before he suspected that detectives were
shadowing his track. This led to his return to Texas, where he

met Bass and the two proceeded to Fort Worth. Here Jim Murphy exchanged $4,000 of the stolen gold for greenbacks. Bass divided the money with his old partner, and that night Davis took the train, and now a score or more of detectives would give much to know where he is.

Early in December Sam Bass, Henry Underwood and Frank Jackson went to San Antonio. Thither they were followed by Sheriff Everheart of Sherman, Tom Gerrin, a rather noted character of Denton, and Tooney Waits, a detective who had come from the north to identify the parties connected with the Big Springs robbery. Waits believed that Underwood was connected with that affair, and declared his readiness to swear that he was Tom Nixon. Sheriff Everhart also acted on the same belief, and went to San Antonio for the purpose of arresting Underwood for Nixon. The name of Bass does not figure conspicuously in the controversy which afterwards sprung up in regard to this matter.

Tom Gerren says that he had a warrant in his pocket for the arrest of Underwood on another charge, and that was his object in pursuing him. He says that he knew that Underwood could not be the man who was known in the robbery as Nixon, because Underwood slept at Jim Hall's ranche in Denton county on the 16th or 17th of September and could not, therefore, have been in Nebraska on the 15th.

There can be no doubt that harmonious action and shrewd management would have resulted in the arrest of all the party. But the officers did not act in unison. One or two prostitutes were let into the secret, which helped to mix matters much. In the meantime Capt. Lee Hall was telegraphed to hurry up with his rangers. But Bass was not the man to be captured by any such hesitating methods, and suddenly he and his associates vanished.

The newspaper controversy between Everhart and Gerren in regard to their failure to make the capture has been extensively published and to that public opinion is referred for a settlement of the question as to who played a bad part.

Bass and party soon returned again to Denton where they remained sometime as will be seen further on.

A freebooter, with ten thousand dollars in ready cash at his command, is apt to prove a great demoralizer to any community not steeled in moral integrity. As he passes here and there among his friends and neighbors, with a pocket full of gold pieces which he deals out with a free hand, buying without pricing and loaning without hesitation, he soon becomes such a convenience and desideratum that men of easy morals and scant conscience do not care to see him driven out of the country or lugged off to jail.

This proved true of Bass' stay in Denton county. His gold made him many friends. "He was always so kind and obliging" that they were "ready to do almost anything for him." No greater curse ever befell that county than this stolen gold. It brought reproach to the whole people, ruin to individuals, and sorrow to many homes. It was an evil which the thousands of good people who live there still deeply deplore, an evil, too, which it will require years to eradicate from the young and susceptible mind.

CHAPTER VIII

TEXAS TRAIN ROBBERIES

Allen Station Robbery—"Your Money Or Your Brains"—Pistol Practice at Short Range and Wild Aim—Going Through the Express Car—Capture of One of the Band—His Trial and Conviction—While There Is Law There Is Hope.

We come now to one of the most daring series of train robberies which ever disgraced this country. The deeds of the old highwaymen, who used to stop unwary travelers at some lonely place in the road and rifle their pockets pale into utter insignificance before the high-handed acts of these modern bandits who dare step upon the iron track of commerce, stop the rushing engine, plunder express and mail cars, while the officers stand pale and trembling before the muzzle of a cocked revolver, and a whole train of terrified passengers sit shivering in their seats until the bold transactioin is over. For outrageous audacity and cool and deliberate proceedings the Texas robberies have never been surpassed not even in the notorious carer of the James and Younger brothers, nor in the bold assaults made upon Union Pacific trains. Blow after blow was struck, even when it was known that the officers on all trains were on the alert, and that all the express and mail cars were guarded by heavily armed men.

The first of these robberies was committed at Allen Station, a very small place on the Houston & Texas Central Railroad, six miles south of McKinney and twenty-four miles north of Dallas. This robbery occured between 9 and 10 o'clock on the night of February 22nd. When the South bound train arrived at the station it was immediatley boarded by four masked men, one of whom leaped upon the engine and in a twinkling had the engineer and fireman under the influence of a cocked revolver. The other

members of the band made a rush for the express car and attempted to enter it, but were repulsed by Mr. J. L. A. Thomas, the
messenger. Mr. Thomas says that he had some express matter for
the agent at that place, and was standing in the door of the car
when the train stopped. The masked men ordered him to throw
up his hands, crying out, "Your money or your brains." He
jumped back in the car and drew his pistol. The robbers then
began firing, he returned the fire, discharging his revolver three or
four times. The robbers fired several shots and then sprang into
the car, previously threatening to burn it if Thomas did not surrender. The bell rope was then cut, the express car uncoupled
from the rest of the train and the engineer was ordered to draw it
over the switch, when the safe was rifled of its contents. The
amount secured was sad at the time to be $2500. It is known now
that it was nearly $3,000.

There was a large number of passengers on board, but they
considered prudence the better part of valor and made haste to
stow away whatever valuables they had upon their persons, all
momentarily expecting to see the robbers coming through the train.
But as soon as they had finished the express car they made their
escape, moving off in a westerly direction.

This bold deed produced much excitement all along the line of
the road. Texas had seen much stage robbing and many deeds of
violence, but this was a kind of lawlessness to which the people had
not become habituated and did not care to see successfully inaug-
urated. But still not much effort was made to capture the robbers, except by the officers of the Texas Express Company. They
at once institutd a vigorous pursuit, and on February 27th captured Tom Spotswood at what he called his cattle ranche on Little
Elm Creek, in Denton county. This arrest was effected under the
leadership of Mr. W. K. Cornish, agent at Dallas, and Mr. Thomas,
the messenger.

Spotswood was taken to McKinney where he had a preliminary
examination, and in default of $2,500 bail, he was remanded to
jail to await his trial at the next term of the District Court.

The trial began in the latter part of June and ended July 2nd.
Mr. Thomas, the express messenger, was the principal witness
against Spotswood. He testified that one of the men who entered
the car on the night of the robbery was not masked, and this man
he recognized as Tom Spotswood, the prisoner. He said that he
had ample opportunity to see him, as Spotswood held a revolver in
his face, while the other men robbed the safe. He noted his pe-
culiar appearance, especially his glass eye. It is something of a

question how straight a man can see with a cocked revolver in his face, but the evidence had great weight with the jury.

Mr. Newman, a saloon keeper at Allen station, testified that Spotswood visited his saloon the day before the robbery and asked him whether there was any gaming done in town, and said that he was a sporting man. He also asked at what hour the train came from the North.

An attempt was made to prove an alibi. Bill Spotswood, brother of the prisoner, and another man testified that Tom slept at the house of the former on the night of the robbery. But two other witnesses testified that they met Bll Spotswood and his companion in the woods the next morning, where they were chopping wood, and they said they had not slept at home the night before becauses they couldln't get across the creek. The jury returned a verdict of guilty, and the prisoner was sentenced to ten years in the penitentiary. But since then he has obtained a new trial. If he lives long enough to wear out all the continuances which the laws of the State permit, the delays which the lawyers ask, and gets safely over the frequent "reversing and remanding" of higher courts, he may go back to his cattle ranche, or he may join some of- the boys at Huntsville. But these are things which "no fellow can find out" until he lives long enough.

CHAPTER IX

HUTCHINS AND EAGLE FORD.

Plan of Attack—Printers to the Front—Brave Express Messenger —Facing the Bullets—A Slim Haul—Great Excitement—The Coolest Robbery on Record—A Lost Opportunity—Another Financial Failure.

Nearly a month passed quietly away after the Allen robbery and the public begàn to feel that the capture of Spotswood, who was regarded as the chief of the band, had put an end to the desperate business. But on the night of March 18th the whole country was startled by the intelligence which flashed over the wires, that still another train had been successfully captured and robbed. This act was also committed on the Houston & Texas Central, at a small station named Hutchins, ten miles south of Dallas. The train selected for the attack was again the southbound through express and mail train from Chicago and St. Louis, which passed the station about 10 o'clock at night.

The following account of the robbery appeared in one of the daily papers the next day:

"The robbers understood their business well, had evidently planned the assault deliberately, and the manner of its execution was prompt and effective. They first took into their possession the railroad agent at Hutchins and a negro, then the engineer and fireman of the train. They also captured two tramp printers from Dallas, who were stealing a ride on the front of the locomotive, and added them to the crowd.

"This squad they marched in front of them to the express car door, so that, should the messenger on board the car fire, the discharge would take effect not on the robbers but on the innocent agent, negro, fireman and engineer, or puncture the valuable epidermis of the newspaper fraternity. The messenger barred the car doors and extinguished the lights, but the robbers soon burst asunder the door. The messenger then fired into the mob, with what effect is not known, but the fire was returned and the messenger wounded in the face. One of the printers also received a wound in one of his limbs, which for the present will operate as a serious check to his perambulatory tendencies.

"Messenger Thomas, being wounded and seeing the futility of attempting any further resistance, surrendered to the mob. The safe was rifled of its contents and the mail car ransacked for whatever plunder the robbers saw fit to appropriate. In regard to the amount of money obtained in the express car there are several rumors. One is that they only obtained a small amount, the express messenger having secreted the bulk of money and valuables in the stove while the lights were out. Another rumor is to the effect that they obtained several thousand dollars. Messenger Thomas continued on his route as far as Corsicana, where he stopped off on account of his wound.

"Mr. Thomas is a brother of the agent who was in charge of the car that was robbed some time ago at Allen station. (He was afterwards rewarded by the company for his bravery.)

"Word of the robbery was dispatched in all directions, but up to noon today no trace of the daring scoundrels had been obtained. Marshal Morton, of this city, with several members of the police force, kept a lookout all last night, but their watching resulted in nothing satisfactory. The marshal rode about thirty-five miles last night, taking in Hutchins and the adjacent country.

"The passengers on board the train were not molested. The robbers, it is said, after transacting their business, took off toward Trinity Bottom, but efforts to track them in that direction for any great distance failed.

"Later reports confirm the statement that the train robbers secured but a small amount of money, probably not over $300."

The great drought prevailing at that time made it impossible to follow the trail of the robbers, and active search for them was soon given up. It was wholly unknown at the time who they were, whence they had come or whither they went.

Matters were now beginning to wear a serious aspect. The repetition of such acts was bringing disgrace upon the State, the traveling public was becoming alarmed, while the express and railroad companies were put to heavy expense to protect their property. As no clue to the robbers had been obtained, and as no great effort was made by the State authorities to ferret them out and effect their capture, it was feared that they might at any time strike another blow.

These fears were soon realized. This time the blow fell upon the Texas & Pacific Railroad, at Eagle Ford, a small station six miles west of Dallas. This robbery was one of the best planned and most cooly perpetrated crimes ever committed. It is almost impossible to believe that the masked men who moved about the train as deliberately as employes of the road engaged in the ordinary operations of the track, were actually robbers coolly plundering cars in the presence of the train men, the express company's guards and a score or so of passengers.

The attack was made on the night of April 4th, and was reported as follows the next day:

"The western bound train was robbed last night at 11:30, at Eagle Ford, by four masked men. The train from the east, on the Texas & Pacific Railroad, passed through Dallas last night a few minutes after 11 o'clock. It arrived at Eagle Ford, six miles west of Dallas, about thirty minutes past eleven. As the depot agent for the Express Company, came out of his office he saw a man come round the corner of the depot upon the platform with a pistol in his hand closely followed by two other men. The first one presented his pistol at the agent and kept him quiet until the other two arrested the engineer and fireman, whom they brought round to where the agent and first man were standing. The fourth man was placed near the passenger coach with a view, it is supposed, to prevent the approach of the passengers. The agent, engineer and fireman were then placed in front of the express door, two of the robbers (one standing at either end of the men under guard), covering them with their pistols, while the leader ordered the local agent to ask the messenger to open the door.

"The messenger refused to open the door when the leader took a stick of wood and broke the door in. On the express car—so we are informed by Col. C. T. Campbell, the Superintendent of the Express Company—were the messenger and a man lately hired as guard. The guard had a shot-gun and the messenger his pistol. Neither made any attempt to fight so far as we can learn. They were both ordered out and into line with the other prisoners. Then the leader, with the express messenger following, entered the car and the safe was opened. The amount taken from the express will not exceed fifty dollars. The mail car was also robbed of several registered packages, but the amount received from this source we are unable to ascertain.

"We are indebted to Col. Campbell for the information above. We also met Mr. Ely, who checks baggage on the train, who informed us that, observing the delay, he went out with the conductor, when they were arrested and held under guard by one of the robbers. He says when the robbers left they retreated with their guns cocked and presented and facing the parties, in readiness for an attack. They went in a northwesterly direction."

Had the guard and messenger made good use of their weapons on this occasion, two of the robbers, at least, would never have rode away in the darkness again.

CHAPTER X

ROBBERS' FASTNESS

Light Breaking—The Hiding Place Discovered—Thick Forests and Sympathizing Friends—Dallas Detectives in Denton—Asleep in the Woods—The Bandits Challenge Their Pursuers to Come Out and Fight—Robbers' Pranks—Getting the Drop on an Officer.

The Eagle Ford robbery greatly increased public excitement and aroused an intense determination to capture the hidden bandits at all hazards.

As yet the State authorities had taken no action except to offer a reward of five hundred dollars each for the capture of the guilty parties. Stimulated by the hope of securing this reward and the desire to rid the country of such dangerous outlaws, a few private individuals, whose names will appear hereafter, made an earnest effort to follow the robbers to their hiding place.

Great effort was also made by the officers of the Texas Express Company to ferret out the bandits. This effort was attended with such success that a few days after the Eagle Ford robbery, they be-

lieved themselves in possession of a full knowledge, not only of
the whereabouts of the robbers, but also of their names. But they
were also convinced that their capture by any means at their hand,
or by civil process, would be well nigh impossible and that the aid
of the State administration must be called into requisition. They
determined, therefore, to appeal at once to the Governor for assistance and to afford a proper justification to public opinion for the
movement, they called upon the managing editor of the Dallas
Commercial and requested a full publication of all the facts which
they had discovered and the conclusions to which they had come
after a long and expensive investigation. The request was com-
plied with and the article at once appeared, on Tuesday following
the Eagle Ford robbery. As this article threw much light on a
question then wrapped in doubt and mystery and turned public attention to the hiding place of the robbers, and as its statements
have been proven almost absolutely true by subsequent developments, we give it below:

ROBBERS' FASTNESS.

Hiding Place of the Train Robbers Discovered.—The Band is Located
in Denton County—Local Authorities Powerless—Nothing But An
Armed Force Can Break Up the Gang.

"After a most thorough investigation of all the circumstances attending the late railroad robberies and a careful following up of
every clue, the detectives are fully convinced that the band of robbers who perpetrated these daring deeds are located in Denton county. The direction taken by the band in the three successive robberies establish the correctness of this conclusion, beyond a doubt.
The band which robbed the train at Allen moved off in the direction
of Denton county, and shortly afterwards Spotswood, who is now
believed by all the detectives to have been the leader of the gang,
was arrested there.

"When the train was waylaid at Hutchins, the roads were so dry
and hard that it was impossible to track the gang, but the indications
were that they also went in the same direction. But the next morning after the affair at Eagle Ford a hot track was found and the
maskers were followed directly to their present hiding place.

"The fastness in which these highwaymen have each time successfully taken refuge is an extensive tract of woodland, full of undergrowth and very difficult of ingress. It is described as a place where
a man could live for a year and nobody ever see him. This forest
contains many log cabins standing among the trees and in such iso-

lated places that nothing but a long search can discover them. It is believed that there are many good people among the inhabitants, but fear of their desperate neighbors compels them to keep their lips closed.

"But many of the people are thought to be more or less in sympathy with the gang. Their houses are always open to them, and when compelled to stay in the woods, they carry them meals and act as spies for them. A detective fully acquainted with the character of the people says there are women among them who would ride fifty miles in a night to warn one of the gang of approaching danger. It is said, too, that they have couriers scattered through all the neighboring country, who keep them constantly informed of every movement of the authorities. Some of these couriers are supposed to be here in Dallas, and constantly act as spies to gather and report the sentiments of the people.

"They are also believed to have regular lines and stations extending as far west as Palo Pinto and northwest to the Indian Territory.

"As is well known, Bass, Underwood and Jackson, who were implicated in the Union Pacific robbery, live in Denton county Underwood alias Nixon, was arrested some time since and taken to the scene of the robbery for trial. But two weeks ago he made his escape and is now at home again. Not long since this gang went into the town of Denton, and hearing that the authorities were trying to capture them, they retired to a mill near the town and sent word to the officers and the whole town of Denton to come out and take them.

"To show the difficulty of capturing the desperadoes in their present hiding place by a posse of civil officers, it is related that a few days since two or three citizens were passing through the woods and upon firing a pistol they heard an answering shot, and repairing to the spot they suddenly found themselves confronted by several men who stood behind trees with Winchester rifles leveled at them and ready to pull the trigger.

"The authorities of Denton county confess their inability to capture the gang.

"There is now no doubt that all the robberies were perpetrated by the same gang. At Allen there were five robbers, at Hutchins and Eagle Ford there were but four—Spotswood having been arrested in the meantime. It is also fully believed that the same parties perpetrated the stage robberies and other deeds of daring in western counties.

"In each case they were traced in the direction of the Denton county rendezvous. It is to be feared, too, that they will next make raids upon neighboring cities, probably going through a

bank or two, and startling the whole country with their bold burglaries.

"The Texas Express and Railroad Companies have spent large sums of money in ferreting out the robbers, and the duty now devolves upon the State to arrest or break up the gang.

"Detectives can do no more, for they have traced the robbers to their hiding place, and can almost name the gulty parties. The local authorities are powerless to capture the robbers, therefore the matter should at once be taken in hand by our State authorities and a sufficient force should be sent into Denton to arrest the guilty parties or drive them out of the country. The companies are now compelled to keep a heavily armed guard on all leading trains. This, of course, involves them in a very heavy expense. The additional expense of the Express Company alone is said to be a hundred dollars per day. As these companies are engaged in legitimate business, they should be protected by the State, no matter what the expense. The state of Texas must protect all its commercial business or it might as well quit. If desperadoes defy local authorities, then the State police might be called upon to assist in the enforcement of law. The name of the State is suffering greatly from these repeated and daring robberies.

"The news of each successful attempt flashes at once to every town and city in the country, and another black mark is scored against Texas.

"Every good citizen deplores this, and we believe that the Governor would be upheld by public sentiment in a determined effort by means of the State police to break up this desperate gang."

This article was extensively quoted by State exchange and republished by different papers throughout the West, and it did much to direct attention to the guilty parties.

The parties mentioned above as meeting the desperadoes in the woods were private detectives from Dallas. The history of the adventures of this squad in Denton country is as follows: The next morning after Eagle Ford robbery Samuel Finley, June Peake, James Curry and one other from Dallas, struck a hot trail leading northward from the railroad, and followed it to the Cross Timbers in Denton country. They continued through the timbers towards Denton, when within about three miles of that place, near the farm of Capt. B. H. Hopkins, they suddenly came upon two men asleep in the woods. They had ridden past them, when some one discovered the men and their horses which were picketed near by. The men immediately sprang to their feet and fired at the approaching party. The fire was returned by James Curry, but no one was hurt on either side. A parley then ensued. The detectives did not feel

sure of their men, and were afraid to shoot them down for fear of killing innocent parties. While they were maneuvering with a view to ascertaining who the men were, the two saddled their horses, sprang upon them, raised a whoop and dashed through the woods. The Dallas party supposing that the remainders of the gang were near did not pursue them, but sent for more force.

Continuing on to Denton, they stopped at the Lacy House on Saturday afternoon, (once the home of Bass) while there the notorious Sam Bass and a number of his associates appeared on the outskirts of the town and, according to one statement, rode into the city. They had heard that the Dallas party were looking for outlaws, and were anxious to know if they were the men whom they sought; if so, they would like to have them come out and try and take them. Messengers galloped back and forth between the excited and defiant crowd and their friends in the city. Finally, later in the evening Bass and company sent a messenger to the Dallas men to inform them, "that they would remain in sight of them for two hours and a half, and challenged them to come out and fight. They stood near the residence of John S. Lovejoy, Jr., (we quote from the Denton Monitor of a few days late) in the eastern suburbs of the city, plain to view from the public square. More than a hundred men saw them."

But as the Bass party outnumbered the Dallas squad, they did not think it best to attempt their arrest without more assistance. Whether this assistance was tendered them by the officials of Denton is a question which we do not care to discuss. It is evident, however, that Mr. Geo. Smith, City Marshall, made an attempt to raise a posse and go to the mill, where Bass had lodged his men and was breathing out defiance to every man from Dallas, but failing in this, he started to the mill alone, and was afterwards followed by a few others. He soon returned with the report that Bass and party had skedaddled. Sheriff Eagan also offered to go out with a posse if the detectives would loan him their weapons, which they declined to do for the reason that they thought they had particular need of them themselves. It was also stated afterwards by the city paper that assistance would have been freely given to make the arrests, if the detectives had furnished the proper papers. But this was impossible, as they had obtained such description of the robbers at Eagle Ford as could be given by the agent and train men, and then followed the trail, hoping to capture the men who answered the description, the names of the robbers being at this time unknown to them. As Sheriff Eagan afterwards spent many weary days and nights endeavoring to capture these same men, there can be no doubt that if he had felt convinced that they were the right parties,

he would freely have offered his assistance to make the arrest. He has the reputation of a brave and efficient officer and does not hesitate to do his duty when he sees it clearly.

But it is certain that the citizens of Denton were greatly in the dark at this time in regard to the guilt of the men, who were thus audaciously defying pursuit in their very midst. That Bass was one of the Union Pacific robbers was generally suspected, and to very many well known. For he had covertly admitted as much to many of his old neighbors. Why he was not arrested, especially when so great a reward was offered for him, is inexplicable. But there was a stubborn unwillingness to believe that Bass and his company had any part in the Texas train robberies. Even after the above developments had been made and published, we find the Denton Monitor, in its issue of April 13th, declaring: "There is no charge against any of this party, in Denton county, except Henry Underwood.. That is for carrying a pistol, and it is not believed he can be convicted on evidence. And it is not believed here that any of this party participated in the train robbery at Eagle Ford, at Allen or at Hutchins. Certain it is that they were here on Thursday night of last week when the Eagle Ford train robbery occurred.

An explanation of this last statement will be found further on, in the sketch of Jackson's career.

The status of the band and the condition of affairs in Denton at this time, is well described by a correspondent who writes in a letter, dated Denton, April 11th:

"Reports about the railway robbers are so numerous and complicated that it would take a Philadelphia lawyer to get at the bottom facts; so I can only give what I've heard, and leave the reader to draw his own conclusions. Sam Bass is the reported leader of the squad now. He is accused of the Nebraska train robbery and of every other one that has occurred since.

"I have seen but one man who might be a robber, and he is said since to have been Bass, and he was certainly well prepared for a fight when I met him (having a Spencer, two Colts and a knife, and well mounted), about four miles south of Denton, last Saturday morning. He approached and very politely asked me 'if I had met any armed men?' and I told him no. He then said 'some fellows had tried to steal his horse that morning and he was after them,' and rode off. I was since told that it was Bass, and that he that day joined his party and went to the mill near Denton, and sent word to the party from Dallas, and all concerned, who were after him, that he was there and to come and take him.

"The range of the Bass party in Denton and the region that abounds with their friends, I hear, lies between Pecan and Cooper

creeks, running east and west across the Cross Timbers, a distance of tcn miles long, between Hickory and Elm, and about four or five miles wide. It is thickly timbered, and they can go over the whole distance and camp most anywhere without being seen, except by friends. They do not confine themselves to this range, but go into the town of Denton frequently, at night, to play ten-pins, drink, and have a good time generally—actually getting the 'drop' upon an officer last Friday night and making him—no very hard job— drink with them, and leaving another friend twenty dollars in gold. They have committed no robberies of private property in Denton county, nor do they molest travelers nor any persons they meet on the road, and are reported civil to all. They protest their innocence, but swear they will never be taken alive, and the one I saw looked as if he was 'that sort of a fellow.' A gentleman in Denton told me that either Bass or Underwood sold $1,000 in gold to a citizen not long ago, and said he knew where $8,000 more was buried. Bass is said to own a saloon in Denton. The difficulties the Dallas party have labored under have been numerous. A few men, in a strange country, try to catch desperate men, well mounted and armed to the teeth, and knowing every inch of the ground, and with friends to warn them and furnish them the latest Dallas and other papers regularly, containing the latest movements and designs.

"The good people of Denton have no sympathy with and give no aid to Bass and party, but what can they do, scattered as they are over a thinly settled country? They feel very sore over the hard things said of them by some of the Dallas papers, and would like the writers to come up and put themselves in their places awhile and have their families and then take all out upon a lonely plantation, in the timbers, often miles from a neighbor, and then see if they would like to tackle or to capture Bass and his 'horse marines.' That 'something is wrong in Denmark,' when men accused of crime can ride heavily armed over the country and into towns without being arrested, can send word where they are and soberly defy arrest no one can deny, but that does not cast a reflection upon the law-abiding citizens of Denton county, who are powerless—scattered as they are, and having among them men who will aid and comfort the robbers."

As this correspondent well says, there should be careful discrimination between the law-abiding people of Denton and the lawless characters and disreputable citizens who brought discredit upon the county.

CHAPTER XIII

TEXAS BAND.

Who They Were—Bass The Leader And "Conductor"—Frank Jackson The "Engineer"—A Dead Man's Head Cut Off In Self Defense—Seaburn Barnes—Henry Underwood—Captured For Nixon—Taken To Kearney, Nebraska—Potency Of Stolen Gold—File and Watch Spring—A Six Weeks Job—Out in the Air—Stealing the Judges Horses—Across the Country—Arrival in Bass' Camp—Arkansas Johnson—Young Lady Killed—Not a Jim Crow Thief—Playing Engineer.

Although the name of each particular individual engaged in the different robberies was not fully and definitely known until after the brush at Mesquite, yet it now becomes important to the interest of the narrative to introduce the bandits to the reader without further delay.

Bass was the leader of the gang in all the robberies and was dubbed "conductor" by the squad, because he always went through the Express and mail cars, attending in person to the safes and mail bags, while to

FRANK JACKSON

was assigned the important duty of capturing the engineer and fireman. For this reason he was called the "Engineer." Being a man of reckless daring and cool determination he accomplished his task with remarkable success, as all the robberies well attest.

Jackson is a native of Decatur, Wise county, Texas, and for a time was considered a respectable young fellow. He lived with Dr. Ross, of Denton, a few years. Some eighteen months ago he got into a difficulty with a negro on the prairies, which gave him much notoriety. As he was alone with his victim, there is no account of the tragedy but his own story. He says that the negro stole his horse and when he demanded his return the fellow showed fight, he then shot him down. He at once reported the matter to his friends, telling them that he had "shot a nigger." After talking to them a little while he said he didn't believe that he had killed him and would go and finish the job. He returned and cut the negro's head off.

For this act he was brought before court and was acquitted it is said, on the ground of self defense. But just what danger his life was in while sawing a badly wounded man's head off, it is difficult to see.

After this he ranged around Denton assocating with wild fellows and leading a fast life. He accompanied Bass and Underwood to San Antonio last December and was believed at that time and later to have been connected with the Big Spring robbery. It was currently reported in Denton during the Spring that Bass gave him one hundred dollars per month to go with him on his exploits. But this is not well substantiated.

Jackson is believed to have assisted in all the robberies but that of Eagle Ford. At the requet of Bass he remained in Denton in company with Henry Underwood on the night of April 4th, in order to prevent suspicion and make the people believe that the gang were all at their homes. That the plan succeeded well, we have already seen.

SEABORN BARNES

according to the best authority in the State, was in all the robberies. The same authority says that if there was a white feather in the gang, the plume belonged to Barnes, as his confederates stood in doubt of his courage. Barnes was a native of Tarrant county and of respectable parentage. He turned cow-boy, and then strayed off among "the wild fellows." Naturally enough, he fell in with Bass and his company.

HENRY UNDERWOOD

is generally believed to have been "the brains of the crowd." He came to Denton from Missouri six or seven years ago, bringing with him his family, who lately resided in Wise county. He settled on a farm, but was wild and associated much with Bass during his career as a horse racer. He was once arrested by Tom Gerren for cattle stealing, which made him a sworn enemy of the queer Tom.

Because of his supposed resemblance to Tom Nixon, he was suspected of being the very man who numbered one of the six at Big Spring, and who fled with Berry. Learning that the officers were in pursuit of him, he fled to San Antonio, where, as we have already seen, the officers followed him, but failed to effect his capture on account of a misunderstanding among themselves. After escaping from the Alamo city he returned across the country in company with Bass to his home in Denton, whither he was followed by Sheriff Everhart, of Sherman, and Pinkerton's detectives, and was finally captured during the last week in December. As he was known to be a daring and desperate fellow, his capture was considered a very hazardous undertaking. But learning his exact whereabouts, the officers surrounded the house, closed in on it pistol in hand and covered their man with the deadly

shooting-irons before he could make his escape or offer successful resistance. He was at once taken to Omaha where he was confronted by the detective Leech, of Ogalalla, who was with the robbers the night they divided the gold captured at Big Spring. Leech was in grave dought about the prisoner's identity, and inclined to the belief that they had the wrong man. But Tony Waits, who seems to have engineered the matter, declared positively that Underwood was none other than the notorious Nixon. Underwood admitted that he knew Collins and Heffridge, and that he had recently been with Bass, but stoutly denied that he was Nixon, or that he had anything to do with the robbery.

He was then taken to Ogalalla for identification by several parties who knew the robbers. Their opinion being somewhat against him, he was taken to Kearney, Buffalo county, Neb., and lodged in jail.

Shortly after his incarceration there, Bass, with his usual liberality, sent him a hundred dollars. He gave seven dollars of this money to a discharged prisoner as he left the jail, with the understanding that he was to provide him some means of escape. The fellow was faithful to his promise, and returned with a file. With this and a watch spring he worked upon the hard iron for six weeks, and was at last rewarded for his long labor, by stepping out into the fresh air of night a free man. Arkansas Johnson, who was confined in the same jail, escaped with him.

The two repaired at once to the stables of the district judge, took a pair of his best horses and galloped away towards the south. When morning came they were far out of reach, and after a hard ride of sixteen days they neared Underwood's old tramping ground in Denton. Bass had by some means obtained intelligence of their escape and of their expected arrival. He mounted his horse and rode away to meet them. His search for them proved successful, and on Sunday evening, March 31st, just four days before the Eagle Ford robbery, he conducted them safely into his camp. Here the lucky Underwood was received with the wildest enthusiasm by his old companions in dissipation.

It has been strongly hinted that the capture of Underwood for Nixon was a put up job for the purpose of securing the large reward which was offered for the capture of the fated Berry's more fortunate companion. But there is not much ground for this charge, for the reason that Pinkerton's detectives are not allowed to receive rewards, but are paid per diem. The detectives were apparently persuaded that they had the right man. As for Sheriff Everheart, he simply effected the arrest on the papers presented. That he received a good recompense for his trouble, and justly, too, there is no doubt.

ARKANSAS JOHNSON

was a stray member of the Texas band, coming in towards the last. But as he played an important part in the later history of the gang, he is placed among them.

Of his former life but little is known. According to the best information obtainable by detectives, his true name was John McKeen, his home in Johnson county, Mo. His father lives near Knob Noster, Mo. He was suspected by the detecives of being connected with the Union Pacific robbery, and search was at once instituted for him. He was discovered at Otterville, but made his escape. Afterwards he was traced to his father's house in Johnson county. In the attempt made to arrest him his sister was shot and killed, but he escaped. Afterwards he was captured and thrown into jail at Kearney. His own account of this matter, as given in Sam Bass' camp, was that he was arrested for stealing lumber.

When he first appeared in camp he was received with ill favor by the members of the gang, even Bass himself is said to have had a very poor opinion of him. Some of the boys said he was nothing but a little "Jim Crow thief," and should never be admitted into the tony society of a band engaged in stopping the wheels of commerce and plucking plunder from rich corporations. (When Underwood, and his influence over the bold leader of the bandits was very great, assured him that he was all right, and could be trusted to play his part well.) That Underwood's usual sagacity did not fail him in his estimate of the shabby looking Missourian, was well proven afterwards. The Eagle Ford robbery occurred on the following Thursday night after his arrival in camp, and as Jackson, "the engineer," remained at home, Johnson was elected to fill his place at the engine. This he did with remarkable coolness and success, capturing the engineer and fireman and bringing them around in front the express car in a twinkling.

One of the last regrets of Bass was that he did not follow Johnson's advice in regard to two very important matters. But this belongs to a later chapter.

Briefly summoning up, we find that the Allen Station robbery was committed by Bass, the leader, Spotswood, according to the sworn testimony of the Express messenger (Thomas), Jackson and Barnes.

The Hutchins robbery was committed by Bass, Jackson and Barnes. Green Hill's name has also been mentioned in connection

with this affair.   But others, who were in Bass' camp, say that there
were but three in the gang.

At Eagle Ford were Bass, Arkansas Johnson, Barnes, and one
other who is still unknown to the authorities.

———————————————◈———————————————

CHAPTER XII

SPY IN CAMP

The Shadows Falling—Scott and Collins in Denton—Making Plans
—Waiting For "An Excitement"—The Mesquite Robbery Arranged—Tripped Up By Time—Pipes And Herndon.

At the time of the advent of Underwood and Johnson in the
camp the bandits were quartered at Bob Murphy's, fourteen miles
beyond Denton, on the road toward Bolivar.   Here they slept in
the barn at night and remained in the woods behind the field during the day, where the time was whiled away playing cards, plan-
ning future robberies or rehearsing old adventures.

On Saturday evening, the day before the arrival of Underwood, just as the sun was sinking behind the trees and the shadows
were falling heavily upon the greensward, two young men entered
the camp, one of whom was to prove the evil hand of destiny to
the band.   This was Will Scott, of Dallas, a young man highly
connected in that city.   The other was William Collins, a brother
of the fated Joel Colins, who was killed at Buffalo Station.

Will Scott came among them as a spy, to effect the capture of
Bass.   As he furnished the greater part of the evidence by which
some of the guilty parties were afterwards brought to justice, we
give his story substantially as it was related in his sworn testimony
during the late trial at Austin.

Knowing the large reward which had been offered for the capture of Bass for his connection with the Union Pacific robbery,
young Scott conceived the idea of effecting his capture through
strategy.   In casting about for the means of effecting his purpose,
it occurred to him that something might be accomplished through
William Collins, because of the relationship formerly existing between Joel Collins and Bass.   He at once repaired to Collins'
house and there learned that some correspondence had already
taken place between Collins and the Jackson's.   He found, too,
that Collins also had a scheme of his own in view, the gist of which

was that he wanted to bring Bass and his companions down to
Dallas county in order to avenge himself on some Duck Creek
farmers who had prosecuted a number of the young men in the
neighborhood for disturbing a merry party of dancers, driving all
the young ladies out of the house and smashing two or three of
the young gentlemen's heads.   The affair had acquired much notoriety, as it led to a libel suit with the Dallas Commercial.

The two at once started for Denton county, not knowing the
exact whereabouts of Bass, but believing that he could be found.
At Denton they obtained information which led them to believe
that he was at Bob Murphy's.   A letter to Murphy was secured
from a Denton lawyer, and they proceeded on their journey.
When they arrived at the camp they found Bass absent in search
of Underwood.   But Jackson was there and they were received
without disturb.   •

Bass returned the next evening and various plans for new en-
terprises were talked over between him and the new comers. Scott
proposed the robbery of a Dallas bank, and a plan to rob a bank
at Weatherford was also considered.

The band then left Murphy's and rode down towards Denton.
While on the way some parties were seen coming from the opposite direction.   Barnes at once dropped at the rear and lagged
behind. "Are you not afraid to meet people in this way?" said
one of the new comers to the robber chief. "Oh, no," replied
Bass, "but Barnes back there always gets uneasy and wants to get
out of the way."

The next day various plans were again discussed and finally
Bass told them that when they heard of another excitement (mean-
ing another robbery) they should return and he would go into some
operation with them.

Collins and Scott left them a few miles below Denton and returned to their homes.

The next Thursday night the Eagle Ford robbery was committed, and as soon as Collins heard of it he started for Denton.
As we have already seen, Bass and his men were galloping around
Denton on Saturday.   Collins must have found them some place
in that vicinity. On Sunday a plan for another robbery was discussed and it was finally agreed to try still another train, this
time at Mesquite, a small station on the Texas & Pacific Railroad,
a few miles east of Dallas.

During Monday and Monday night the party found their way
to William Collins' house, situated in Dallas county, some twelve
or fourteen miles east of the city of Dallas. On Tuesday young
Scott set out again for a visit to Collins, to see what he was going

to do, now that "the excitement" had been heard of. He found
that Collins had already been to Denton and that the whole band
were right there.

It was now Tuesday evening, and that night was set for the
robbery. Collins had been to Mesquite that day and pronounced
everything all right. Bass' party consisted of himself, Jackson,
Barnes, Underwood and Arkansas Johnson. To these had been
added Sam Pipes and Albert Herndon, two yong men who had
lived in the neighborhood for some two weeks, working upon the
farm by day and having a wild time at night. They both figured
in the assault upon the dancing party.

After nightfall the party mounted their horses and set out for
Mesquite, William Collins accompanying them. But as the train
was late, and as they reached the station a little after the regular
time for it to pass, they thought it had already gone by and returned
to Collins' place. Here they concealed themselves during Wednesday. When night came, a parley ensued as to who should go, Bass
objecting to so large a party. He said that the booty was likely to
be small at best and would not reward a large crowd. He also
sa.d he would rather have his new friends act as outside men; that
they could do more good in that capacity than by going under
fire. William Collins mounted his horse, but finally yielded to the
chief's persuasion and got down. Henry Collins was all the time
averse to having anything to do with the affair, and besought Pipes
to the last not to go, telling him that that night would not end the
matter; that a day might come when this expedition would prove a
sad affair. But Pipes was "train struck" and would listen to no
reason. It is said that he formerly lived in Missouri near the rendezvous of the James and Younger brothers and that evil shadows
had fallen across the bright beams of childhood fancy. Be that as
it may, himself and Herndon accompanied the band, making seven
in all. This turn in affairs relieved Scott from all necessity of par-
ticipating in the affair.

CHAPTER XI

MESQUITE.

A Sharp Fight—Brave Conductor—Firing From Underneath The
Train — Convict Guards Empty Their Shot-Guns — Passengers
Flat on the Floor — A Woman Who Wouldn't Hold Up Her
Hands — Determined Resistance of the Messenger — Threatened
With Fire—Car Saturated With Coal Oil—Surrender—Wounded
Robbers—Captain Alvord.

At the close of the last chapter we left the banditti mounted and
ready to ride away under cover of darkness to meet the evening
train. The plan of the robbery seems to have been the same as the
one previously executed with so much success. But it was greatly
disconcerted by the bravery of the conductor and express messenger. The near presence of a convict contruction train, surrounded
by several guards armed with double-barreled shotguns, also added
much to the confusion. The last robbery really proved the only
exciting one of the whole series, and had there been a few more
determined men on board the train, it might have resulted very
disastrously to the reckless bandits. From the reports published
next day and from subsequent developments, we gather the following account of the fight:

"When the hour for the train arrived the robbers stood under
cover of darkness just behind the depot. Soon the roaring sound
of the cars was heard and a few moments later the train was seen
rushing in from the east. The whistle sounded and the locomotive
stood at the depot. Before it had fairly ceased its puffing and
snorting, the cry, 'hold up your hands! hold up your hands!' rang
out upon the air.

"Captain Julius Alvord, the conductor, who was on the sleeper,
had just stepped forward to the front passenger car and on to the
platform. He had his lantern in his hand. He saw some parties
near him who called to him to come to them, cursing him as they did
so. He managed to put out his lantern as soon as possible and stepped on to the car, crossing over to the other side from the depot,
he went imediately to the sleeping car and got a larger pistol. He
had with him a small derringer. He put this in his coat pocket and
took his larger pistol in his hand, went to the rear of the sleeper,
and opened fire on his enemies with the pistol. There were three
parties firing on him, and being too much exposed he went down
the steps off the platform, and at this time he was shot through the
arm and a large hole shot in the back of his hat. He passed between

the cars to the opposite side of the car from the robbers, with them
still firing on him. The engineer here attempted to start the train,
but was stopped, only moving a short space. Conductor Alvord then
took his position under the car and continued his firing. From this
position he made several shots, but his wound became so painful
he came out and re-entered the second-class coach. He found the
passengers flat on the floor. He examined his wound by the dim-
light, and concluded to go to the sleeper and have it bound up.
When he stepped from the second to the first-class car, several
shots were fired at him. He looked to see that no one was near
when he came out, but the robbers, while they were not in sight,
kept shooting. Before entering the sleeper he stuck his head out
and called to them, and asked them what they wanted. They
asked him who he was, and he said, 'a passenger.' They replied,
'we want money.' They cursed him, and fired on him. He then
went into the sleeper. His wound was bleeding profusely, and
he took a sheet and bound up his arm and laid down. There were
in the sleeper two gentlemen and their wives, going to Fort Worth.
One of the gentlemen had considerable mony. He had concealed
it in different part of the car, thinking if they found part of it
he would still have some left. He was dressing; was very cool;
he said: 'Conductor, I have a pistol, can I do anything? If you
say so, I will go out and try them.' Captain Alvord told him to
remain there and shoot them if they tried to come in. There were
about twenty passengers on board, and seven of them were ladies.
No one else offered assistance. Captain Alvord thought he hit one
of the men with his second shot, as he fell back suddenly to a pile
of lumber immediately after he fired. The porter also heard them
talking at the rear of the sleeper, and thought one was shot. The
porter also heard hem speak of Conductor Alvord, saying, 'He is
a brave fellow, it would be too bad to kill him.'

"Mr. D. J. Healey, at the time clerk at the Windsor Hotel, in
Dallas, had quite a little experience, and his story will not be without interest: He left the city at 5:10 going east, and went up
to Terrell, where he met the western-bound train. When near
Mesquite he said he stepped on the platform so as to gain as much
time as possible. He wanted to see the agent at Mesquite, who
was a personal friend of his. He says he was the first man on
the platform; that he got off before the train stopped. As he
stepped down he saw a man step on the platform a short distance
off, who was soon followed by eight or ten more. He started forward and the first man made for him, while the others started for
the express car. The man who came to h'm presented his pistol,
and told him to come on, and to throw up his hands. He took in

the situation at once, and told him that he was not armed. He did not follow very rapidly, and the fellow kept curs'ng him. The robber went backwards, and he followed the pistol. The engineer made an attempt to start the train; Healey's guard became excited and started to assist the others to stop it, hallooing 'don't let them get away.' He took advantage of the absence and took $100 out of his vest pocket and put it in his boot. When he started to raise up from doing this the robber came on him with pistol in hand and again ordered him to follow. He started along slowly, keeping his eyes on the man in front, when some one came behind him and struck him on the side of the head with what he supposed was a pistol. It stunned him a little, and as he revived he told the fellow he was a coward. He looked round but could not tell who struck him. He then watched his chance to get away. His guard soon gave him an opportunity by leaving him a few paces, going toward the express car.

He whirled and ran, and the robber fired; he ran east; he had run a short distance when he looked back and found he was still pursued; the robber again fired, and then returned to the train. He ran along the road until he came to the construction train where the convicts were; he hid under the train and stayed there about twenty minutes, when he approached the guards on the construction train and told them what was up. There were eleven guards on the construction train, and one went down the track and encountered a picket from the robbers' force. The robber fired and the guard returned the fire. Healey said he wanted to get to the train but could not tell when to approach; he finally started, but before he reached the train it pulled out and he was left. While lying under the construction train he heard the sound of horses' feet, at the same time the robbers were still firing around the train. As the guards who had charge of the convicts could not leave them they fired a number of shots from their posts, and with some effect, as we shall see hereafter.

In the meantime one of the robbers had, as usual, taken charge of the engineer and fireman with a cocked revolver. Another subdued the station agent by the same means, but a woman who lived at the station, proved a much more refractory subject. In spite of all comands to hold up her hands, to stop, etc., she ran away to her room and locked herself in. "Conductor" Bass and two of his confederates repaired at once to the Express car to "administer" on its effects. But Mr. Curley, the messenger, was a very determined man and faithful employe, and thought he could take much better care of the company's property than Conductor Bass or any of his trusty fellows. When commanded to open the

door he positively refused to obey. The bandits told him "that if he didn't open it they would break it in." He told them to "go ahead." But as the shrewd Bass was well aware that the brave Curley and a heavily armed guard stood within with cocked revolvers in their hands, and would have nothing to .do but pull the trigger the moment he showed his head, he thought it wise not to execute his threat. He then shouted to them that if they didn't draw the bolts and swing back the door, he would set fire to the car and burn them up. They replied that they wouldn't do it. Then a can of oil, which had been hidden at a convenient place under the platform, was brought out and the car saturated with oil. Bass informed the messenger what he had done, and said he would give him just two minutes to surrender. Knowing the desperate character of the men, Mr. Curley concluded to surrender. The door was shoved back, and the robbers entered the car. After a hasty and almost fruitless search they retired. The mail car was also visited and a few registered letters were taken from the bags. The whole amount of money secured made but twenty-three dollars apiece for the seven robbers.

The bandits retreated to the horses which were hitched nearby, and mounted and rode away.

The rapid firing had awakened some of the citizens living near by, and one or two approached and fired at the assailants, but no one thought it wise to attempt to pursue them.

Afterwards it was learned that three of the robbers were wounded in the fight. Pipes received a shot in the side, which proved a very damaging fact against him afterwards. Underwood and Barnes were both wounded in the limbs, but not seriously. The former hurried away to Denton, but the latter is said to have gone to William Collins' place, where he lay for a day or two concealed in a hay-stack.

Captain Alvord continued on to Dallas, where he was taken to the Windsor Hotel and at once received medical treatment. It was found that the ulna of the left arm was badly shattered. The wound proved verp painful but not dangerous. On examination of his hat, it was found that a very close call had been made for his head, as a large piece had been shot out of the back part of the hat.

Captain Alvord is a single man about thirty-five years old, and was born in New York, leaving there when about seventeen years of age. Before the war he was connected with the O. & M. R. R., and also the Hannibal and St. Joe. He entered the army in the 30th Illinois regiment as a private, was afterwards promoted and when mustered out was Adjutant. Since the war six years of his

life was spent with the M., K. & T. R. R., and for the past two years he has been on the T. & P. R. R. He is a man of great nerve, and has been much praised for his heroic defense of his train. Mr. Curley was also very highly commended for his bravery.

This was the last of the Texas train robberies, and we come now to another turn in the history.

CHAPTER XIV

GATHERING THEM IN.

Public Feeling at A High Pitch—Loud Calls To The Governor—
Major Jones on the Ground—Detectives Hurring Up—Company
Of Rangers Organized—Arrest Of Pipes And Herndon—Billy
Collins Going To Swear Them Out—His Arrest—Attempt To
Betray Bass—Plan To Rob A Dallas Bank.

It would be impossible to describe the excitement and indignation which this fourth robbery produced.

The next morning as soon as the news spread over Dallas intense feeling was manifested and it continued to increase as the day advanced. Men walked the streets with stern faces and clenching their fists declared that this thing must be stopped if it took the whole State of Texas to do it. It was felt that local authorities had been somewhat remiss in the performance of duty or were totally unable to cope with the gigantic proportions which the evil was assuming. Loud calls were made for more determined action on the part of the Governor. That official had written not long before to the railroad officers, in answer to an inquiry from them, as follows:

"Be assured I will hereafter, as heretofore, offer in proper cases suitable rewards for the capture and conviction of all such criminals. Whatever power the law gives to the executive will be promptly exercised in aid of yourself and the civil authorities, towards providing against the recurrence to robberies of our railway trains, and to secure the speedy arrest and punishment of the felons who perpetrate them."

He was now vehemently urged by the State press to come forward with whatever aid lay in his power.

Major Jones, commander of the State police, was at once sent to Dallas to institute a vigorous search for the robbers. The city was also full of detectives, while sheriffs, constables and policemen were flying about in every direction. Every few days some poor fellow, who happened to have the smell of powder on his clothes or a wild look in the corner of his eye, was gobbled up and brought to town. But he always turned out to be the wrong man.

Major Jones proceeded to organize a small company of mounted police, or rangers, consisting of thirty picked men. These were sworn in on May 18th and placed in command of Lieutenant June Peak, formerly Recorder of Dallas.

In the mean time the Major had fallen in with Will Scott, who told him that he had been making some effort to decoy Bass into a corner where he might be captured, and that he had obtained some valuable information. This information was then imparted to him and thus the commander of the State forces was placed in possession of all the facts relating to the robberies, the idenity of the guilty parties and their hiding places. These facts have already been given in the narrative.

As it was known that Pipes and Herndon were still in the county and that frequently they came to the city, steps were immediately taken to arrest them.

A plan was arranged that Scott should go out to the Collins place and reconnoitre. If the two were there he would remain over night and Major Jones could come out and effect the capture, if not he would return to camp. As he did not return, Major Jones went out and found Scott and Henry Collins and Pipes at Mr. Collins' house.

They were all put under arrest. Scott then informed him that Albert Herndon was at Mr. T. J. Jackson's, some few miles away. A negro arrested on the outside of the house also gave the same information. Scott and Henry Collins were then turned loose, and Herndon was arrested and the two were brought to town. They very emphatically denied their guilt and seemed indifferent to proceedings. After a preliminary examination they were admitted to bail which was promptly furnished and they were turned loose again.

About this time, John and Morris Griffin, on trial for robbing the Express Company at Paris of ten thousand dollars, were found guilty, but only sentenced to two years' imprisonment.

Major Jones learning this fact concluded to change the case against Pipes and Herndon to the Federal District Court. New papers were accordingly taken out and the young men were again arrested and had a hearing before U. S. Commissioner Fearn. They were also taken to the jail and searched, as it had become known that one of them had been wounded. No marks were found upon Herndon, but a small scab was discovered on Pipes' left side. When questioned in regard to it he said that it was a little boil, then he admitted that it was a pistol shot wound which he had received from one of his comrades in the country, but he had concealed the matter to keep the man out of trouble. He became somewhat confused in his story, looked despondent and it was plain to the officers that he would be glad indeed to get his shirt back over that scab.

Commissioner Fearn, held the parties to appear before the Federal Judge at the next term of court, the bonds being fixed at $15,000 each. This they were soon prepared to give. But U. S. Marshall Russell, who had taken charge of them, received a telegram from Judge Duval, stating that he had sent a bench warrant to him and that the prisoners must be produced at Tyler. At eleven o'clock that night they were ironed together and put aboard a special train and hurried away.

This no doubt saved some of their friends a heavy forfeiture, as they would probably have jumped their bonds, had they regained their liberty. This was April 26th, sixteen days after the Mesquite robbery.

April 29th, William Collins was attached as a witness to appear at Tyler. That afternoon he appeared in one of the newspaper offices "to get the thing right in the papers." He was interviewed at some length.

The substance of his statement will be found in the following calloquy:

Collins.—Pipes and Herndon can easily established their innocence by proving an alibi. It is true that they have been pursuing no calling for some time, but have been idle.

Reporter.—Mr. Collins, this as you know, is urged against these two young men; can you tell me anything concerning it?

Collins.—They have been idle for the reason that they were making preparations with myself to go west with me, where I intended to take a drove of cattle. Pipes, who owned land in the county has sold it, and Herndon has leased his farm.

Reporter.—Where did these young men make their homes?

Collins.—Pipes has been living for some months at my father's residence, about a mile from where I live in the White Rock neighborhood, and Herndon has made his home at my house, staying there most of his time.

Reporter.— You say they can establish an alibi on the night of the Mesquite train robbery.. Will you tell me what facts you feel justified in making known as to their whereabouts on that night and the nature of the proof they can make?

Collins.—Pipes and Hernon were both at my house that night. They went to bed about ten o'clock, and I and my wife will both establish the fact that they both staid at my house up to that hour. I saw them go to bed myself in a room above my own from which they could not have left during the night without passing through my bed-room. The next morning early they were still in their room, and I saw them both get up. There are persons who work on my place by whom they can prove that they were at my house that night.

Reporter.—I suppose there is no doubt that Pipes has about his person a gun-shot wound received at that time?

Collins.—Oh, yes! Pipes has a wound but it was received before the Mesquite train robbery. I know all about it. It was an accident and it was made with a little old pistol which is at my house now. That matter will be satisfactorily explained and made perfectly clear at the proper time.

Reporter.—If these young men can so easily establish their innocence—in other words, if there is so little grounds for their arrest, what do you suppose has led to their being suspected and arrested?

Collins.—Well, I don't know, but suppose persons at enmity with them have put up a job on them for the purpose of injuring and harassing them.

The following description of Collins appeared with the report of this interview:

He is a young man, apparently between twenty-five and thirty years of age; is tall and well formed; has a frank, open, honest-looking face, a clear grey eye, high forehead, dark hair, and is very intelligent, conversing well and at perfect ease.

It was remarked after the handsome but unfortunate young man passed out of the room, that unless Marshall Russell and Major Jones were much less shrewd than they have credit for being Collins would soon be behind the bars with his old friends. This remark proved correct, as Collins found an indictment waiting for him at Tyler, and he was at once put under arrest.

An indictment was also found against Henry Collins, but he has not yet been captured, as he became alarmed after the arrest of his brother and joined the Bass gang.

In the mean time Scott Mayes, a saloon keeper at Denton, and a negro named Scaggs of the same place, had also been arrested and taken to Tyler. They were charged with being accessories to the fact.

Soon after this, May 2nd, Bob Murphy, a cattle man in Denton, and Green Hill, a sporting character, were arrested at their homes in Denton and taken directly to Tyler. They were charged with being accessories to the robberies.

On the same day Sheriff Everheart, of Grayson county, arrested Henderson Murphy, father of Bob Murphy, Jim Murphy and Monroe Hill. They were all charged with being accessories and also with harboring Bass and his party. It was alleged that about a week before the arrest, the gang had a frolic at old man Murphy's house, where Underwood's wife was boarding; that they cut up "high jinks" and had a "high old time," practicing with their pistols and boasting what they would do with the rangers.

Just after the arrest of Pipes and Herndon, Scott determined to make another visit to Bass' camp. After consultation with Major Jones he set out for Denton and succeeded in finding Bass at Green Hill's, about six miles below Denton. He informed him of the arrest of Pipes and Herndon, but nevertheless found Bass willing to enter into a plan to rob a Dallas bank. The exchange Bank, of Gaston and Thomas, was agreed upon as the proper one to "go through." (It is not on a record that Bass made any mention of the State Saving Bank.) The plan was partly mapped out but left in an indefinite shape, Will Scott returning to Dallas to complete the arrangements.

But in the meantime Bill Collins had become suspicious of Scott and sent a letter to Bass by Mayes, telling him that Scott was a spy, and exhorting him to hang him to the "handiest" bush. Scott was anxious, however, to return, but Major Jones, with more mature judgment, saw the danger and forbade his going. The Major was also averse to the plan to rob the bank, as he knew that the bandits were a desperate set of fellows and in the confusion of the moment some harm might be done to persons employed at the bank, and that the robbers might succeed in capturing the contents of the safe and escaping. In other words he did not care to be held liable for damage to property with so poor a prospect of success.

CHAPTER XV

THE GREAT CAMPAIGN AGAINST BASS.

Swarms of Pursuers on the Trail — The Ball Opened — Driving
Them Through The Woods—The Air Filled With Random Bullets
—A Red-Hot Breakfast—Escape To The Western Mountains—
The Game Sprung Again—Two Days' Fight—Four Foolish Farm-
ers Who Turned Robber Hunters And Got Captured.

The great summer campaign against Bass which continued until
the last of June, was opened about April 24th. More men were
employed in this campaign, more powder burned, more bullets buried
in post oaks and green hillsides, more horses rode to death, more
ground galloped over, more false alarms given, more prophecies
blown into thin air, more expectations blasted and fewer men captured than ever before occurred in any similar campaign in human
history. Of course we do not include in this statement the prisoners
who quietly submitted to arrest at their homes, making no attempt
to escape.

About the date mentioned above, Bass' old tramping ground in
Denton county began to swarm with rangers, detectives, sheriffs, and
the now excited citizens of the county. On that day Sheriff Eagan
had a brush about a mile and a half west of Denton with the now
bold, bad man, who in better days gathered with his family at his
own fireside. Nothing was accomplished by the skirmish, and nothing more was attempted until early the next week. Then Sheriff
Eagan again took the war-path followed by a host of deputies and
all the excited citizens who could find a gun, borrow a horse or
seize the steed of some unfortunate farmer who happened to be in
town. Peak also arrived in Elm Fork bottom at the same time with
his company of rangers. It is said by an eye witness that there were
fully 150 men on the trail. The Denton people felt the strictures
which had been made upon their county and were determined to rid
it of the bad gang who had so long defied arrest.

A correspondent who witnessed the scene gave the following
description of it:

"They had them corralled once or twice, but Captain Bass & Co.
did not seem to care for the fight much, and depended on the run
for it. Many shots were fired. The 'flying couriers,' as Bill Arp
used to call them, were out in full force and made their Texas ponies'
hearts sick. Every ten minutes one would come dashing up to the
court house at full break neck speed to report progress. One would

say 'they had sent him back because his horse was broken down;' but all agreed that 'they'd have Bass sure before nightfall.'

"Your correspondent took an excursion into the 'bottoms.' Flanking the 'flying couriers' he saw some rare shooting of Winchester's and Spencers at random. He caught a glimpse of Captain Bass and his party a little distance off. Bass sat on his horse like a Comanche Indian, and didn't seem to care a continental for hurting any of the pursuing party. He and Underwood left the horses they were riding and took it afoot. Judge Hogg captured Bass' horse; but down into those everlasting, hidden bottoms went Bass & Co. In this skirmish Deputy Marshall Minor, of Denton, had a rare tumble from and with his horse. They rolled over and over and looked like fourteen gentlemen in one. Sergeant Minor returned to Denton to report a sprained ankle.

"Judge Hogg, Capt F. E. Piner and Capt. T. Daugherty were out after the desperadoes, as were all the prominent citizens. The saloon keepers, who were accused of harboring and encouraging the Bass party, were also out in force, and if they could have been captured in these terrible bottoms, I believe it would have been done.

" The people of Denton are as hospitable, as kind, and as law-abiding a people as I have ever met in Texas. The reason Bass gives for coming to Denton is a sensible one and a good excuse for his leaving here. Dallas, Collins, Ellis, Tarrant and other of our counties are high and rolling, with but few hiding places, but when you strike the cross timbers up here and get into Elm and Hickory bottoms, you see four or five Chickahominy swamps all boiled down into one. The foliage is dense—the vines hang in masses and the undergrowth thick, and it is not good daylight until 12, noon. Now, Bass, Underwood and company, knew these 'bottoms' thoroughly. For nearly a year, and before that, they have made every 'hog path' a study, and they knew them as well as 'Marion and his men' knew the swamps of the Pedee in South Carolina. That they would naturally resort to such a section in their trouble, any one who has been through them can see at a glance. The section lying between Elm, Hickory, Pecan, Copper, and Little Elm, surrounding Denton city north, southeast and west, with the Cross Timbers and the grand prairie to hide and run in, affords a place for the operations of a small force of armed and desperate horsemen, such as is not to be found elsewhere in the United States.

"That no one was killed is easily accounted for. A lot of green soldiers were after men trained to the bottoms, and to hardship. Firing with Winchester rifles or Spencers from the hip, on a horse galloping through the dense timbers of the bottoms, is not likely to hurt any body much. Bass and party knew this, and though it is said Jackson's ear was shot off, the only wounded men I saw were Sergeant Minor, whose horse fell on him, and a young Mr. Hart, who placed his Winchester upon the toe of his boot and shot one of his toes off.

"Of course, all engaged had wonderful stories to tell. 'How terribly they rode to the front, and all that, but the truth is, 'Captain Bass and his Horse Marines'—5 or 7 to 150, 'got away with them,' after three days riding around and firing— and as 'Charely,' of the Denton Monitor says, 'they're lit out and lit." That they have gone all the good people here rejoice to know, and so will every one be glad if they meet with their deserts. That they are bold, desperate, determined men, no one who has seen their ways can doubt. They have sworn 'to die before being taken alive,' and every one who knows the men think they mean what they say. But let justice be done to Denton county and her good people, and hereafter, 'let him who is without sin cast the first stone."

The correspondent was right at that time in saying that the band had fled from the country. The three days' fighting alluded to occurred April 30th and May 1st and 2nd. On Wednesday morning, May 1st, Lieutenant Peak came upon Bass while he was preparing to eat his breakfast, in the woods about four miles southeast of Denton. The bandits had barely time to escape, leaving their food on the fire. One horse was captured. They at once turned their course towards the west, closely pursued by the rangers. On Thursday they were heard of in Wise county and on Thursday night Peak arrived with his command at Decatur, the county seat of Wise. On Saturday, May 4th, a telegram was received from Decatur, saying that Peak had divided his command and was scouring West Fork bottom.

But Bass and his men being well acquainted with the country, made good their escape and nothing worse was heard of them for some days.

Sheriff Everheart, who with his usual alacrity and courage had been actively engaged in the pursuit, returned to Sherman May 17th and reported that he did not know where Bass was, though he thought he was in Denton or Wise county. He said the whole country, through Denton and Wise counties, was alive with scouting parties out after Bass, and that his operations were greatly retarded by these bands of zealous but inexperienced robber hunters. On two different occasions his squad was charged and captured by companies of citizens. He said also that there were bands of men pretending to hunt Bass who were really his accomplices and keeping him posted in regard to the movements of the troops.

Sheriff Eagan was also heard from about the same time and reported that he was in Montague county and believed that Bass was making for the Indian Territory.

It was a general belief at that time that he was trying to reach the Territory or the Nation. Certain it was that he had vanished in the tangle wood of the western counties, and nothing more was heard from him for something over two weeks. The public began to think that the gang had surely enough escaped from the State. But suddenly the following telegram was flashed across the wires:

Griffin, Texas, May 18.

"Sam Bass with five of his men is surrounded on Big Caddo Creek by Berry Meadows, sheriff of Stevens county. Meadows was re-enforced by ten men from Palo Pinto last night at 2 o'clock. He expected to make the attack at daylight this morning. Some fighting was done yesterday and the day before. No damage done on our side. It is not known whether any outlaws were hurt."

The manner in which they were discovered and the incidents of the attempted capture, were furnished by a correspondent of the Fort Worth Democrat as follows:

"Deputy Sheriff Freeman was informed last week by a woman of the neighborhood, near Caddo Creek, that parties answering to the description of the train robbers were there. He, with one ranger, and Messrs. Amis and Paschall of this town, went into that section to ascertain something more dfinite, and learned that Bass, Underwood, Jaskson, Barnes, and two others, supposed to be Welch and Collins, (Henry Colins had joined the band some time previous to this), had been camped there in the mountains for upward of two weeks. A brother-in-law of Jackson, and several other kin and friends are living near Caddo Creek, and had furnished them with supplies. They are reported to be flush with twenty dollar gold pieces, and from events developed more recently, they are found to have numerous friends in that vicinity. Having gathered the desired information, the ranger reported to his camp in Shackelford county, and the balance repaired to Breckenridge, where Sheriff Meadows and Deputies Freeman and Hood selected several picked men, and on Sunday started for the scene of action. At midnight they sent back for reinforcements, and twenty old shot guns were collected together and the same number of volunteers. Before all of these new recruits arrived, the sheriff's posse came upon the gang near the store, thirteen miles east of here, on the Palo Pinto road, and an engagement ensued, in which about forty shots were fired by each party, and at one time three of the party dismounted and fought from behind trees. It is thought one of their horses was wounded. They afterwards chased the robbers about two miles into the mountains. As the gang was so much better armed than the Sheriff's party, and were acquainted with the locality of the mountain defiles, they then had little to fear. On Monday night they camped among the trees and thickets near Taylor's store, and the sheriff's party on the prairie one-third mile distant.

"Tuesday morning, May 26th, the Sheriff and his posse were gladdened by the arrival of the gallant rangers from Shackleford county, nineteen in number, armed to the teeth, and their force had also been increased by Deputy Sheriff Owen and eight picked men from Palo Pinto town. The rangers were under command of Lieut. Campbell and Sergeant Jack Smith, and the Breckenridge party of fearless Deputy Sheriff Freeman. Sergeant Smith, of the rangers, stated that if they could find them, they would capture the robbers dead or alive, if they lost half their men in the attempt. On Tuesday they followed their trail through mountains gaps and defiles, and among the hills and valleys in their winding course, but up to twelve o'clock last night had not overtaken them, though the gang had come back to near the starting point. At McClasen's store, four miles further east, they purchased eight dollars' worth of provisions, and left word for the pursuers that they would stand their ground and give them a desperate fight, and that they did not propose to be bull-dozed, all of which is supposed to be a blind, and that they in reality were preparing to strike out for parts unknown. It was ascertained that they had been trying to swap off one of their horses. They are said to be well mounted and each armed with a Winchester rifle and a pair of six shooters. Before the arrival of the rangers the Sheriff had summoned four or five citizens in that neighborhood to secure arms and join his posse.

"The Bass gang passed the same party soon after, before they had obtained arms, marched them down to the store and treated to bottle beer. It is said that parties in that vicinity have carried the Bass gang baskets of provisions and kept them informed of the movement of their pursuers. One of the gang, it is reported, is suffering from a wound received in Denton county. One of them remarked to some person at the store that they were no petty thieves, that they interfered with no private citizen, but holding out a handful of twenty-dollar gold pieces, 'that is what the Sheriff and his posse want.'

"They are said to have $5,000 with them and to have buried the balance. In getting volunteers from Breckenridge, it was quite manifest that a greater portion of the citizens considered it their duty to join the home guard and gallantly paraded the streets in their vigilance to find Bass whom they proposed to demolish forthwith.

"The rangers from Coleman county are expected across the country, to intercept them in case of a retreat in that direction.

Additional parties from Griffin passed here last night to join the forces and aid in the capture.

But all efforts to surround them proved unavailing and a few days later, May 31st, it was telegraphed from Breckenridge that the Sheriff's posse and rangers had given up the chase after Sam Bass and party and that they left Bass boss of the situation in the cedar brakes and mountains, fifteen miles east of that place, where they easily eluded their pursuers.

During this "drive" after Bass the following amusing incident is reported to have occurred. We give it as told in one of the daily papers:

"A gentleman from the· vicinity of the late scenes of the attempted capture of the Bass and Underwood gang, tells the following unexampled story in connection with a fruitless effort on the part of four gallant farmers who were bent on heading the robbers off and taking to themselves the glory and consequent profit of their capture.

"The rumor spread like wild-fire through the neighborhood that Sam Bass and his confederates were scouring through that part of Shackleford county, brazen-faced and publicly proclaiming they did not give a continent damaged darning needle whether school kept or not.

"Our gentleman informant, Mr. Nance, of Young county, says that the robbers rode up to a store located near the edge of Shackleford and Young counties. purchased some provisions and after leaving a note stating that they were the robbers and were going on south-west to Taylor's store several miles beyond, paid for what they got and gently rode off in the direction indicated in the note. As quick as possible the alarm was given, and one of the Deputy Sheriffs of the county, accompanied by four farmers armed with shot-guns, started in hot pursuit and overtooked the Bass gang moving leisurely along the highway.

"How to capture them was now indeed an enigma. They finally decided to separate, the Sheriff to remain behind and intercept a possible attempt to retrace their steps, and the four shot-gun heroes were to move rapidly forward in a circuit, come into the road suddenly on their side, and with cocked guns order their rich game to surrender or suffer the alternative—death. The scheme wore the aspect of a plausible one, and the four farmers started to execute it.

"Putting spurs to their horses, they shot off to the right of the road and were not long in getting in an obscure position by the road-side ahead of the robbers, where they could not be seen until so desired. Apparently unconcerned and careless, the four robbers drew near, and as they got opposite, the stern demand to halt and surrender was given.

"Just then a wild whoop from behind proceeded from the woods, startled the shot gun heroes, and in the disorder which ensued, the robbers getting possession of their arms, got the drop on their would-be captors and turned the table by 'taking them into the fold.'

"The whooping party was one of the scouts who had been following the farmers and perpetrated the successful trick. All four of the pursuers were taken in charge, marched to Taylor's store, and in less than half an hour were boiling drunk through the hospitable treatment at the hand of their captors, who left them shortly after accomplishing their aim, in possession of a note cautioning them to make no such absurd attempt to bull-doze a gang of Sam Bass' train robbers."

CHAPTER XVI

THE GREAT CAMPAIGN AGAINST BASS CONTINUED.

Bass Returns to His Old Stamping Ground—Dashes Into Denton and Recaptures His Horses — Long and Hot Pursuit — Pursurers' Horses Killed and Riders Wounded—Murphy Joins the Band— Running Fight Across the Prairie — Bandits Surprise at Salt Creek—Desperate Fight—Arkansas Johnson Killed—End of the Campaign—Why It Failed.

After the escape of the band in the vicinity of Brackenridge, nothing more was heard of them until May 5th.

On the evening of that day a courier suddenly dashed into the little town of Elizabeth, Denton county, and as he reigned in his foaming steed, shouted out in breathless haste that Captain Sam · Bass was again on his old stamping ground, having just been seen in the neighborhood of Mr. Burnett's farm, on Denton creek, about nine miles distant.

A posse was immediately collected and started in pursuit. But Bass was riding hard to accomplish something very different from an escape, and soon left farmer Burnet's place far to the rear.

The next morning, just as the first rays of light leaped over the eastern woods, the dashing train catcher and his band galloped down the streets of Denton past the houses still closed and silent, meeting here and there an early riser who gazed in astonishment at the dust-covered riders. Reaching the center of town they halted their weary steeds in front of Work's livery stable. Here they found Charles McDonald, an employe, who had just risen and begun his stable work.

He was immediately informed that they had come for the three horses captured from the band by Sheriff Eagan on the 1st of May, and that he must "bring them horses right out." At the same time Bass had drawn close to him and was flourishing a revolver in close proximity to his face. He refused to get the horses when Jackson, who had also drawn near, began to strike him over the head with his pistol. He then commanded McDonald to stand still and sent Jackson and Carter, a man who had lately joined the band, into the stable to saddle the horses and bring them out. This was soon done and the band put spurs to their horses and dashed away.

Two men, heavily armed, were sleeping in the upper story of the stable at the time. They heard the noise below, one of them siezed a double barreled shot gun, looked out of the window and saw Bass and his men starting away. He looked at them as they rode across the green into the street, but reserved his ammunition for a better opportunity.

The men engaged in this daring raid were Bass, Jackson, Underwood, Barnes, Arkansas Johnson and Carter. Carter is a reputed cattle thief, and though he had but lately joined the band, yet he had for some time been more or less connected with them. It is not known whether Henry Collins was in this raid or not.

Before the clatter of the retreating horses hoofs had died away, the cry of "Bass," rang along the streets, and soon the whole town was in the wildest commotion. Horses were hurried from the stables, bridled and saddled and in less than half an hour a whole troop of riders dashed off in the direction taken by the bandits.

To give a full account of this pursuit of the gang across their old tramping ground, from the time that they were seen near farmer Burnett's until they disappeared near Boliver, we introduce the details as furnished by a correspondent at the time. The letter was dated at Elizabeth and was written by a young man engaged in the pursuit. He says:

"On Wednesday last about 4 o'clock p. m., a special courier arrived in Elizabeth bringing the message to Capt. Withers, deputy sheriff, that Bass and company had made their appearance in the neighborhood of Mr. Burnett's, on Denton creek, about nine miles northwest of this place. Capt. Withers immediately summoned a posses to his assistance and started in pursuit, arriving where they were last seen about dark, and scoured the bottoms, which proved fruitless, the night being too dark to strike any trail. They dispatched a man at once to Denton and the utter darkness prevented his arriving there until late the next morning.

"At daylight Capt. W.'s crowd struck the trail, which they followed direct to Denton. Arriving at Denton they learned that Bass and company had arrived there a little before daylight, taken from Work's livery stable the two horses which had been captured from him during his last trial, and went to the residence of Sheriff Eagan, roused him from his pleasant slumbers and told him to get up, "as the country was full of thieves," and then rode out of Denton. Strange to say, but yet true, a man stood on the loft of the stable, with a double-barreled shot-gun and a belt of cartridges, and let Bass proceed with the horses.

"Withers and command, after stopping a few seconds to hear the tale, proceed on their trail and rounded them up at Pilot Knob, about six miles south of Denton, on Saturday morning, when they had the first battle. One of the pursuers was wounded. This was George Smith, the Denton marshal, who offered to arrest Bass at the mill for the Dallas men. Bass then fled.

"Runners were sent out for recruits, one going to Denton and the other starting to his place, but he was overtaken by Bass, ordered to dismount, and they took his saddle, cut his bridle into strings, turned his mule loose, and told him if they caught him again they would kill him.

"They then rode off towards the timbers. But before they reached the trees, Captain Withers was up and made a charge, but Bass was too well armed and Withers had to dismount. After he dismounted Bass and company poured about twenty-five volleys right at him, cutting the ground all around him and throwing the dirt all over him. Bass then retreated, taking a south-easterly course.

"Up to this time not over seven were in close pursuit, and Bass having seven in his outfit, and armed with long-ranged rifles.

"From this place Bass started in the direction of Alton, thence to Ballard's Mill, thence to Davenport's Mills, pursued closely by the command, which now had swelled to nearly forty. At Davenport's Mills Bass stopped, bought some coffee-pots and inquired for ammunition. It was discovered that one of Bass' men was shot in the side. They then took for Medlin School House, where Withers again came up within shooting distance and gave them a round, but inffectively. They then took to the timber and rounded and came back very nearly on the same track they went down, struck out by Doe Harris' and through the timber again, when the trail was lost about ten o'clock at night. The men being worn

out, they went to camp. The next morning at daylight they were
again in the saddle and on the trail. After proceeding a mile and
a half they came to where Bass had camped. The trail being fresh,
they made good time, and came upon Bass and company cooking
their breakfast, about a mile and a half from Ballard's Mill. Bass
opened the fire, shooting down two of Withers' boys ponies. The
volley was returned and one of Bass' men was shot. They then fled,
putting the wounded man on his horse and one of the other getting
up behind to hold him on. One of Bass' horses, his cooking utensils
and provisions were captured.

"Your reporter arrived at the battle-field about an hour after
the battle and joined the boys, who then trailed Bass to the
McKinney road, about six miles east of Denton, where the trail
was lost about five o'clock Sunday afternoon, and we returned.
This nearly accurately describes the whole movement.

"Bass is well mounted, well armed and has plenty of money
and legions of confederates in the Cross-Timbers, and it may con-
sume a week before he is finally either shot down or captured, as
Withers and Eagan and their men are determined to follow as
long as any trail can be found.

"It is said one of Bass' men has fallen out.

"Bass has all the advantage in the world, as he has nothing to
to do but to ride through the woods, while the other party have
to often spend hours hunting up the trail.

"Sunday evening two gallons of whiskey, some tobacco and
provisions were shipped to Elm bottom from Denton, and it is
thought it went to Bass, as he has runners in all directions.

"He has made the remark that he intends to kill Capt. Withers
before he leaves the country, as he believes Withers is the only
officer in the c ountry who is seeking to arrest him.

"From what we saw of the men's performance on Sunday
they are all determined to have the whole outfit or die trying.

"The whole party are very much fatigued, and their horses are
jaded to the last.

"The people of Denton are very much aroused and are fully
awaked to their duty, and are rallying to the front as fast as
they can.

"We left about fifty men in pursuit Sunday evening, and no
doubt the crowd is doubled by this time. Courtright and two
others from Fort Worth caught up with the crowd shortly after
the battle Sunday morning, and took the lead with Withers at
once.

"The men from this place who have been and are still in pursuit, are Jack Bates, G. M. Powell, J. Goff, M. Kinser, W. H. Berbe,

A. E. Allen, and Ed. Dunning, and are as brave and true as ever breathed.

"Tuesday, June 11.—The news came here late yesterday that the trail had been entirely lost in Elm bottom, and another that Bass and company had passed Bolivar with fifteen men strong. The informer said Eagan and his posses had returned to Denton, and Withers and his men were still in pursuit. We can hear most anything, but we are positive that Withers has not lost or been off Bass' trail one hour since he struck it on Wednesday last.

"If they did pass Bolivar, it is almost sure that they are making for Lost Valley, where they will be joined by about thirty other outlaws, who have been traced to that point.

"It may seem to your readers, Mr. Editor, that in four battles, and in the last not over fifty yards apart, that Bass surely ought to have been taken in ere this, but if you were better acquainted with some of his dodges and tricks it would appear different.

"All the creeks had to be swum up to Sunday morning, which made pursuit troublesome.

"In the fight at Pilot Knob, Riley Wetzel was shot through the calf of the leg by one of Eagan's men. He was conveyed back to Denton by Mr. Allen.

"Later.—Bass went into Bolivar Monday, bought a sack of flour, four sacks of coffe and clothes. He was remounted and in good trim. They ate their dinner near Bolivar. The crowd cannot be far in the rear, and no doubt will run him into Lost Valley. A company of rangers will be up by this evening. One Murphy joined Bass near Bolivar."

This was Jim Murphy, a man who figures very prominently in the narrative before it closes.

In the fight at Ballard's mentioned above, John Work, a son of Captain Work, of Dallas, was wounded in the shoulder, and the firm had two horses killed.

On the evening of June 10th, the same day on which Bass was reported to have gone into Bolivar, the band was engaged by a party under Sheriff Everheart in a running fight on the paririe, in the north part of Denton county.

A farmer who witnessed the fight, says that when he saw them the Bass crowd was riding seven abreast and going at a furious gait. The Sheriff's party were from three to four hundred yards behind, stringing out single file, and spurring their horses to their utmost capacity. Some thirty or forty shots were fired. About three or four miles farther on the Bass crowd reached a strip of timber, dismounted and disappeared.

The next seen of the gang was in Wise county, where they were

discovered by Captain Peake and his command. The Rangers had
followed them into the tangle wood along Salt creek, near Cotton-
dale, on the morning of the 12th. Most of the day was spent in
searching through the dense bushes for the hidden robbers. In the
afternoon they were suddenly discovered lounging under the trees
on the bank of the stream, their horses having been lariated out a
few rods away. The Rangers at once crossed, opened fire upon them,
and the fight which ensued was the sharpest and most successful of
the campaign. The bandits seemed more anxious to escape to their
horses than to fight, but the Rangers crossed the stream and headed
them off from the horses. During the fight, one of the robbers exposed his person boldly to view, Sergeant Floyd, a crack shot of the
Stonewall Greys, of Dallas, seeing him, dropped upon his right knee
and taking deliberate aim fired, killing him instantly. The victim
afterwards proved to be Arkansas Johnson. Bass was standing at
his side, but escaped unhurt.

Shorty after the fall of Johnson, the robbers slipped away through
the tangle wood and escaping under the bank of the creek, concealed
themselves in a large excavation. While here, as was afterwards
learned, one of the Rangers came very near and stood plainly in
view. Jackson leveled his rifle upon him and asked Bass if he
should shoot. The chief said, "no, not unless he turns this way."

Underwood, however, escaped to the horses, mounted one of them
and was returning, when he was met by Captain Peak in the woods.
A pistol duel at once opened, but Captain Peak shouted to his men
and Underwood put spurs to his horse and fled. The heads of the
robbers' horses were now seen in a clump of trees not far distant.
As Captain Peake did not know where the robbers were, and fearing
that they might be with the horses, he ordered his men to fire upon
the horses, which they did and two of them were killed. The rest
were captured.

Johnson was buried by citizens living near the place. .   ..

Thus ended the career of one of the bravest and most hardened
of the gang. Though despised at first, he soon proved bold and bad
enough for the worst undertakings of the band. Johnson showed
more wisdom than any of his confederates in his estimate of Will
Scott. As soon as he took a good look at him, he pronounced him
a spy, and advised Bass to get rid of him. Bass afterwards greatly
regretted that he did not follow the advice. One other piece of
advice given by Johnson, was, to leave Texas and go to Arkansas
and rob a bank. Bass also regretted that he did not take this advice.

During this Salt creek fight Henry Underwood, Henry Collins and
Carter disappeared.. Underwood was never seen again by Bass.

The next night after the battle the robbers stole some horses in the neighborhood and, as has since been learned, directed their course towards their old hiding places in Denton and Cooke counties. But they were not discovered again by their pursuers.

The Salt Creek fight, which occurred June 12th, virtually ended the great campaign against Bass. It began with a light skirmish April 24th, and continued seven weeks. During that time Bass and his men were almost continually harassed by their pursuers. They were kept on the run day and night, through the woods, across prairies, over swollen streams (for unprecedented rainfalls occurred during a greater part of the time), on horseback and on foot, and they were under fire something like a score of times. But still only one of the band was killed and none captured.

If the question be asked, why the campaign proved so fruitless, it should be answered because of the forests and the tangle wood. This gave the robbers all the advantage. They could hide in a moment, and could at any time turn upon their pursuers and shoot them down. That they did not kill them can only be accounted for on the supposition that Bass was determined not to stain his hands with blood if possible to escape without it. He said afterwards, that he frequently lay in the woods within ten feet of his pursuers, but allowed them to pass unmolested.

Had he so determined, it is easy to see that he could soon have made it so fatal to pursuing parties that but few would have had the courage to follow his trail through the dense forests.

One of the most experienced detectives in the country gives it as his opinion that the tramping ground selected by Bass, is one of the best places in the country for such purposes. In Nebraska, where the Union Pacific robbery was committed, and in Kansas where Collins was captured, the country is open, and a single horseman can be seen for many miles away. It is difficult to pass in any direction without being discovered. But in Denton and adjoining counties, parties can ride through the forests for days without being noticed.

It must also be remembered that numerous confederates in all parts of the circuit constantly afforded invaluable assistance, supplying the robbers with provisions and conveying secret intelligence at the slightest note of alarm.

The campaign proves, that in such a country a pell mell drive after robbers, is not the best way to capture them. Brains are better than eyes. The keen, strategic mind runs further, faster and more surely than the swiftest steed. Robbers are not wise. Their abandoned course shows a perversion and obscuration of intellect as well as of morals.

Again, all robbers are traitors to human society. By the very

principle and practice of their profession they are easily led to betray one another. How well these principles were established before the close of Bass' career, will be seen in the subsequent narrative.

---◇---

CHAPTER XVII

PIPES AND HERNDON

Prisoners at the Bar—Trial Postponed—Taken to Austin—A Sad
   Contrast—The Man Who Killed Another for Snoring—Trial and
   Conviction—Confession—Collins Runs Away.

We return now to the case of Pipes and Herndon and the other
prisoners who had been captured and taken to Tyler to await the
sitting of the Federal court. The bonds were fixed at a very high
amount and the prisoners remained in jail until May 21st, when the
case was called.

Long before the hour for opening court the room was crowded
with spectators eager to catch a glimpse of the men who had dared
to put the brakes upon a train without going through the formality
of a regular engagement with the company.

The case was numbered 1456 and was entitled, "The United
States vs. Sam Bass, et al.

The Prisoners brought to the bar were Samuel Pipes and Albert
Herndon.

District Attorney Evans announced ready for trial. But the
defendants, Judge Barksdale and Sawnie Robertson, of Dallas, and
Robertson & Robertson, of Tyler, felt persuaded that there were
several good reasons for delay..

First they presented a plea in abatement, in substance alleging
that two of the grand jury which preferred the indictment were
disqualified, because they gave aid and comfort to the Confederacy
in the late civil strife.

Mr. Sawnie Robertson presented the motion, and in a forcible
argument insisted that the bill should be quashed. Judge Evans
replied on the part of the government and made a considerable
State's rights speech. Judge Barksdale closed the argument for the
defendant.

The motion was overruled by the Judge.

Next the defendant's counsel made a motion to transfer the case
from the District Court to the Circuit Court of the United States.

This met the same fate at the hands of the Court.

Col. John C. Robertson presented additional exceptions to the indictment, alleging that the description of the offense was insufficient.

This Judge Duval also overruled.

Next the defense asked for a continuance, pleading the absence of material witnesses and moving that the continuance be granted. The Judge replied that there was but one ground in the application on which he would grant the continuance, and that was that Mrs. Shipley, a material witness, was sick and unable to attend. After careful examination the continuance was granted on this ground. The case was therefore postponed until the court met at Austin, June 24th.

The prisoners were at once taken to Austin and placed in jail to await the calling of their case.

That they did not find this prison as pleasant as the free air of North Texas where they used to scamper over the prairies and run scrub races on Sunday and raise a wild whoop along the roads by night, is easily gathered from a brief description of the place given by one of their old neighbors, who, while attending the trial, visited their prison.

Passing through a hall walled in by solid masonry, the jailer unbolted a pair of heavy iron doors and he found himself in a large room filled with rows of iron cages. It was a hot day in July, and a July day in Austin is not to be described by any figure of speech which will not stand the test of white heat. The room was dark and not very well aired. The men were stripped to the waist and the perspiration was dripping from their bodies. The cages were of solid iron bars, the floor was sheeted with iron. There were no bedsteads in the cells, a blanket or quilt answered all sleeping purposes. From one to three occupants were in a cage. There were more than three-score prisoners in all. Among them was the notorious John Wesley Hardin, accused of twenty-seven murders, and captured in Florida last summer, by running him off in a train of cars amid a shower of bullets which laid one of his confederates stark and stiff upon the sand and frightened a score of passengers out of all recollection of themselves.. John wrestled and struggled on the floor while the train ran twenty-five miles and then cried all day because he had been caught.

He was pert and saucy as ever and advanced to the front of his cage for a chat.

"This is a very bad place to come to," said he, "people better keep out of here. They say there is honor among thieves, but don't you believe it. There's not a word of truth in it. When they can't steal from anybody else, they steal from one another." (A remark which had a terrible fulfillment for Bass a few days later.) "They tell lots of lies about me. They say I killed six or seven men for snoring, but it isn't true. I only killed one man for snoring."

But Pipes and Herndon did not have to remain here long, for their trial opened July 2d, and after a few days delay got well under headway. It ended July 17th with a verdict of "guilty of robbing the United States mail and endangering life." The jury affixing the penalty at ninety-nine years in the penitentiary, but the judge sentenced them for life, the difference being somewhat immaterial.

The principal part of the evidence for the prosecution was that already given in Will Scott's statements. He testified to the facts relatel in chapters 12 and 13. The defense attempted to break down his testimony by an effort to prove that he did not relate these facts to Major Jones until after the arrest of the prisoners at the bar. They also attempted to prove an alibi, somewhat according to the plan intended by William Collins. The wife of Collins was present but was not allowed to testify.

After their conviction and especially while on their way to the penitentiary, the condemned men freely admitted their guilt, charging William Collins with getting them into their troubles. They said they were present at Mesquite and assisted in the robbery. They greatly praised Curley for his heroic defense of the express car, but attributed the numerous little gun-shot wounds received that night to the buckshot fired at them by the convict guards.

Scott Mayes, Monroe Hill and Bob Murphy were immediately re-arrested on another charge. Afterwards they gave bonds for their appearance and were set at liberty.

Green Hill and the negro Scaggs are still in jail at Austin.

It is a well established fact that just previous to William Collins' arrest he received eighteen hundred dollars from Bass. It is difficult to discover whether this money was intended for use in the trials of Bass' various friends, or whether, as is claimed by parties cognizant of the transaction, it was given for the purpose of establishing a stock ranche in the West, which was to be used by Bass and his band as a rendezvous.

Collins was taken with the remainder of the prisoners to Austin, but afterwards gave bonds in the sum of $15,000. But he jumped his bonds and the unfortunate sureties have been notified of the forfeiture by the U. S. Marshal. Unless Collins is speedily produced, there will be a nice little bill to pay. Collins' whereabouts since his hasty departure are unknown.

CHAPTER XVIII

THE BETRAYAL.

No Honor Among Thieves—Murphy Plans to Betray Bass for the
Price of Liberty—Bargain Made With the Authorities—Memorandum of the Contract—Public Greatly Excited—Two Report-
ers Who Wanted To See Bass' Mother—Immense Stretch Of
Imagination.

One of the disciples was pronounced a thief and it was prophesied
that he would betray his Master.

John Wesley Hardin, whose native shrewdness and bitter experience make him a good authority on the subject, says there is no
honor among thieves. Bass had many friends who constantly be-
trayed law and justice to shield him. At last it was found that one
these friends was just as ready to betray Bass himself, when the
price was offered That price was not thirty pieces of silver, but
liberty—liberty to go in the sweet sunshine and walk the green
fields of earth.

As we have already seen, Jim Murphy and his father Henderson
were among the prisoners taken to Tyler. As the flying train
hurried Murphy away to prison, he was busily ransacking his brain
for a plan of escape, and even before he crossed the dark threshold
of the jail he half determined upon a sacrifice of the leader of the
bandits who had so often assisted him with his stolen gold.

But as no one can tell the story better than Murphy has told it
himself, and in order to avoid all appearance of injustice to the parties concerned, we give as it came from his own pen. In his statement made at Austin, July 26th, an exact copy which has been
furished us, he writes as follows:

"I will give you all the true statements of the plan that was laid to
catch Sam Bass:

"I, W. Murphy, was arrested May 1st, 1878, by Sheriff Everheart,
of Grayson county, for harboring Sam Bass. I was innocent of the
charge, and told Everheart so. I asked him why he did not tell me
long ago that he wanted Sam Bass. He gave me no answer of any
satisfaction but pushed me off from my family and put me in jail at
Sherman. Walter Johnson took me from the Sherman jail and put
me in jail in Tyler. On the way to the jail at Tyler I hinted the plan
for capturing Sam Bass to Taylor, and he said he would send Johnson to see me soon. Johnson came to see me after I had given bond.
I told him that I could plan a job to capture Sam Bass if I was foot

loose. Johnson told me that he would see me again soon. So he
went off and came back with June Peak, and we talked the matter
over. June says, 'I will go and see Major Jones.'. The Major came
and talked with me about the plan for the capture of Bass. At this
time I made a contract with Major Jones as to what he would do for
me and my father if I would catch Sam Bass. He said if I would lay
the plan for the capture of Sam Bass, that he would have my case
and my father's dismissed, and that he would see that I should have
my part of the reward and his part too. He said that he did not .
want any of the reward, and that I should have what was right. I
worked this plan under three men, Jones, Peak and Johnson. Nobody else was to know anything about it. They were the men I
relied on. After a short time Sheriff Everheart worked into the
secret through Johnson. The first time that Everheart came to me
I gave him no satisfaction. The second time he came a man by
the name of Taylor was with him. Taylor told me that whatever
Everheart told me would be all right with Johnson, and I let him
into the secret against my own will."

The following is the memorandum of the contract entered into
with U. S. Attorney Evans:

"Whereas, James Murphy stands indicted as an accessory in
robbing the United States mails, in several cases now pending in
the United States District Court at Tyler, and, whereas, I believe
public justice will be best subserved, hereby, I, Andrew J. Evans,
United States Attorney for the Western District of Texas, bind the
United States as follows:

"1st. If the said Murphy should leave Tyler I will protect
him and his bondsmen at this term of the court.

2nd. If the said Murphy shall be instrumental in securing
the arrest and delivery to the United States Marshal of the Western District of Texas, of all or any one of the following principals,
in their order (Bass, Jackson, Underwood, Barnes and Johnson)
in said indictments, then all prosecutions are to be dismissed as to
said Murphy, growing out of his acts as accessory to the said principals; to be done upon certificate of Major John B. Jones.

"3rd. In case the said Murphy shall use all reasonable and
possible means in his power to capture the said Bass and his above
named associates, and if Major John B. Jones will certify to such
facts to the United States District Attorney, then the said Murphy
is to have the relief named in sectino2nd above, although he may
be unsuccessful.   (Signed)   A. J. Evans,
     May 21st, 1878.   U. S. Attorney."

This contract, as seen by the date attached, was entered into
May 21st, the first day of the trial at Tyler.

In order to convince Murphy of his sincerity and his power to
secure his liberation from all charges, Major Jones had the case
against his father dismissed at once.   With many admonitions to
prove faithful and perform his task well, he was allowed to go on
his mission.   It was immediately given out that he had "jumped
his bond," and so published in the papers.   Previous to Murphy's
departure it was arranged that he should report progress to Major
Jones, Captain Peak or Deptuy Marshal Johnson, according as it
was most convenient.   It was also the purpose of Major Jones to
keep Captain Peak's command at Dallas, so that Murphy would
know where they were and could make his calculations accordingly.
If word reached Peak that he was needed by Murphy, he was to
hasten to the required point.   If, however, it was more convenient
to communicate with Johnson, the latter was to immediately telegraph to Captain Peak at Dallas.   But on account of the pressure
of public opinion—because the public didn't understand the situation—it was found very difficult to keep Captain Peak's command
at Dallas. . People were greatly excited and extremely nervous lest
Bass should escape.   They were also impatient of the delays and
failures which had attended the efforts of the Sheriffs and rangers
in Denton and the western counties.   The air was constantly full
of rumors, and every straw of Bass news was eagerly caught at by
the reading public.   Newspaper men were constantly on the alert
for the latest, the truest or the wildest report about this "greatest of
modern bandits."   An amusing instance of this occurred to two
newspaper reports in Dallas.

Notwithstanding the fact that the grass had been green upon
the grave of Bass' mother for seventeen years, yet one of these re-
porters was suddenly persuaded one morning that he heard her
name echoed through the air on the voice of the winds.   He immediately laid plains to clutch the fact and make it his own.   But
we leave the story to a reporter who slyly watched proceedings
and told it at the time as follows:

"This morning an item reached the ears of a newspaper man
which was full of promise.   He was told in confidence that Sam
Bass' mother was in town, and the strides he made as he struck out
for her supposed stopping place, the St. Charles hotel, were a
wonder to see.   Here was fruit, surely and he congratulated himself as he tore along at his good luck, and he chucked at the thought
of having struck the biggest thing yet.   In fact, he hadn't been so
happy since the Saving's bank "busted."   The proprietor of the hotel
named was the first man run against, and the news gatherer press-
ed him into service and the two proceeded to find the individual
who could put them on the track of Sammy's mamma.   Their search

was successful, and a sable gentleman with chalk eye and gizzard foot was ushered into the reportorial presence. Interrogatories breathless and pointed were put to Sambo, and imagine the consternation created when he said the venerable Mrs. B. had just departed for Denton.

"Knowing the utter uselessness of trying to overtake a member of the Bass family, the reporter determined to make the most out of what the darkey knew about her, and a scorching examination revealed the fact that the woman told him that she was 'some kin to Massa Bass,' and considering her elderly appearance he 'kalkelated' she was his mother; and as she went towards Denton he 'kalkelated' that town was her destination. This put a different face on matters, indeed. The reporter looked daggers at the 'coon, threatened to stab him through and through with a Faber, and he would have reached the office sadder and wiser if it had not been for the fact that just as he was leaving another victimized newspaper man came stealing around, having gotten wind of the bonanza. 'Misery likes company,' you know, and No. 2 was permitted to work up the case, and together the boys shared the immense water haul."

———————◈———————

CHAPTER XIX

THE BETRAYER WITH THE BAND

Murphy Joins Bass—Plans to Betray Him Fail—The Band Set Out On Their Last Trip—Visit to Dallas County—Muhphy's Treachery Discovered—Jackson Saves Him From Death—Visit to Rockwall, Terrell, and Waco—Searching for a Bank to Rob—The Last Piece of Stolen Gold — Murphy Writes to Jones — Arrival at Round Rock.

The adventures of James Murphy in the prosecution of his plan to capture Bass are of thrilling interest and are given nearly as he afterwards related. As soon as he was released at Tyler, May 21st, he returned to his home near the line between Denton and Cooke counties. At that time Bass and his gang were in the mountains and cedar brakes of Stephens county. Murphy could not reach them at the time. But about two weeks later they returned to Denton county, and on the 6th of May dashed into the town of Denton. They were hotly pursued on the 6th, 7th and 8th and on the 9th they took refuge at Murphy's house. When he saw them coming he is reported to have bounded forward to greet them with unbounded hypocritical enthusiasm. He told them, which was true, that he had been "laying out" for two weeks, trying to get with them, but the opportunity had never offered.

About this time he learned that Sheriff Everhart was also in the secret and he entered into a plan at his own house to betray the band to him. Murphy told him he would have two of them back of his field that night and that he could arrest them if he wanted to. But for some reason which Murphy does not know the sheriff's party did not come. Murphy then mounted a good horse and arming himself with a six-shooter, went off with the band. As we have already seen, they continued their course towards Bolivar and secretly entered the town.

From Bolivar Bass hastened west to Wise county, where the Salt Creek fight occurred. While there Bass stole some horses and returned to the north part of Denton county. Murphy says:

"I laid a plan to bring the gang to Bolivar, and after I got the gang there I told Clay Withers and Taylor just where the gang were, and that they would go to Billy Mount's stable, in Denton, to steal his horses, and I would stay outside. The men I had there were Sam Bass and Frank Jackson. No action was taken that night. The next night three of us, Sam, Frank, and myself, went to Mount's stable and stole a horse, there in Denton, and then went to Elm Bottom. We stole the horse about 12 o'clock at night; Sam and Frank went into the stable and got the horse. We got to Elm Bottom about daylight and slept some fifteen or twenty minutes there. Then we went across Big Elm at Rock Crossing. We camped on this (east) side of Elm Bottom and stopped there for breakfast; laid there until noon; then Sam noticed a good many men on the road and directed us to saddle; then we went along the bottom to near Hilltown, where we camped again; stole some corn and had dinner. We then traveled through a big pasture and got kind of lost. We stopped at the house of a man named Burton, I think, some ten miles from Dallas, all night. Next day we went to W. O. Collins' and stopped there about two hours. Sam went there but could not get anything. We did not go to the house. We then went northeast to a church and met Seaborn Barnes."

While here Henry Collins, in company with a stranger, brought the news from Fort Worth that Jim was a traitor in the camp; that he was in collusion with the Rangers. Upon this Jim was notified that he would have to die, and they asked him what he had to say. Jim replied that it was all true what Collins said, but that he had entered into an agreement to get away from the officers of the law, inasmuch as he was indicted and would have to go to the penitentiary. His intention then as now, was to give them the "slip," and that if they would let him live and remain with them, he would take the lead in all they undertook. Jackson plead for Murphy also. He said he had known him from a boy, and didn't believe that Murphy would betray them. They let the matter rest there. But Bass and Barnes were not convinced that all was right and were sullen all that night. The talk was a long and earnest one, and there is no doubt that the party were on the point of riddling Murphy with bullets. After that Bass kept a strict watch over him, and he found it almost impossible to communicate with any one.

Murphy resumes his narrative of this and succeeding events as follows:

"It was this meeting with Barnes that nearly cost me my life, as on that night the rest of the crowd got the information that I was a spy. There was a stranger with Barnes when we met him. We talked a long time there, and I convinced Frank Jackson that I had sold out to Major Jones to fool him and get out of a bad scrape myself. He stuck to me or I should have been killed that night, and I owe my life to him. After talking a long time we started together and kept on towards Rockwall. The next night we stopped all night near Rockwall. The camp was near the edge of town, and while there Bass looked up and saw the gallows on which a man had recently been hung. He said, 'Boys, if I had seen that I would not have stopped here. It makes me feel bad to look at it. How I would hate to die on that.' (Many readers will remember that this gallows was erected for the execution of Garner, the murderer of the Sheriff. But the night before his execution, his wife went into the cell with poison upon her person, intending to die with him. The poison not proving sufficient, Garner hung his wife to the prison wall with the bale of a bucket, and then choked himself to death by filling his mouth and nostrils with strips of cloth torn from his clothes. When the jailor visited the cell at the dawn of day he had just expired. No greater prison horror has ever occurred in the country.) "We then struck out towards Terrell, and got there late in the evening and struck camp just south of the town about 8 p. m. Next morning Bass and Jackson went into Terrell and viewed the bank, and came back and said they did not believe we could make it. I proposed viewing some other banks and taking the easiest. We then struck out for Kaufman, and Barnes and me went into Kaufman and bought a suit of clothes, and left my old clothes in the store, as a guide to Everheart and others who might be on the hunt for us. We found no suitable bank there and passed on down to Ennis. I could not leave them a moment to telegraph, and had to stay with them. We stopped a day and night at Ennis. Sam and me examined the bank, and thought it unsafe to tackle it, as it was too high for us. We then struck for Waco; reached there about 1 p. m., and camped on the north side of town. Frank and me went into town, looked around, put our horses into a livery stable, got shaved, got dinner in a restaurant, then got a $5 bill changed in the bank and saw lots of money, and we returned and reported to Bass, and I suggested that he (Bass) better go and see it as Frank Jackson was excited. Next morning we were on the south side of town, and Bass and Jackson went and looked at it, and decided to rob the Waco bank. So that evening we all moved our camp up on the west side of the Bosque river, to look out a place to retreat to. Sam then proposed that Frank Jackson and me should go in and see where to hitch our horses and get some bacon, lard and coffee, and arrange for retreat if we robbed the bank. On the way I worked upon Frank as to the dangers of it, so that he decided not to rob it, but I did not know what conclusion they had arrived at until next morning at breakfast. Up to this time I had no word from anybody, and was anxious to get some one on the trail. Bass said. 'Jim, we'll go where you say.' We then went south of town again, and that night, before leaving Waco, Seaborn Barnes went and stole a fine pacing mare, with two white hind feet, the one he had when he was killed.

"At Belton I sold SeabornBarnes' horse and gave a bill of sale in my name, to leave a clue. I also wrote a letter to Johnson and Everheart—a few lines only—for God's sake come at once, as we are bound for Round Rock to rob the bank there. I slipped it in the postoffice. After we left Belton we went to Georgetown. There I wrote to Major Jones, at Austin, that we were at Georgetown, and on our way to Round Rock to rob the railroad bank, or to be killed, and to prevent it for God's sake. I just got that letter in as Bass came in. He asked me what I was doing in there so long. I said I was trying to talk this man out of his paper. The man took the hint, threw down the paper, and said he would loan but said he could not sell it. Bass said 'that's all right,' and I read it to him one side. Then we went on to Round Rock and camped out about a quarter of a mile west of old Round Rock town, on the San Saba road, and bought feed and grub there from May & Black; also bought some in the town. This was on Sunday night. We fooled around until Friday."

Some matters of local interest are not given in the above narrative.

At Terrell, Hall & Company's bank was thought the best one to

rob, but they didn't think it safe to try the job. Their appearance
in this town as afterwards described was as follows:

"One drizzly day, some weeks ago, there rode down Moore avenue five mounted men, with a shotgun each thrown across their
saddles in front of them. The leader was a devil-may-care looking
fellow, with a saucy cock of his sombrero on the side of his head,
and an eye like an eagle. The balance of the cavalcade were rowdy-
ish enough, wearing slop-shop clothes and rakish hats. They disappeared at the east end of the avenue and finally turned up on
Broad street on foot. They were seen to enter Messrs. Holt
Bevins & Cooley's bank, come out and walk up and down the Star
Block, and then go in the direction of Uncle Jim Harris' livery
stable. Back of this, it appears, they had hitched their horses, and
springing into their saddles they rode leisurely in a northeasterly
direction. It now turns out that these men were the famous bandits—Sam Bass and his reckless followers—as the description since
minutely given of them, corresponds to a dot to the noted and chiv-
alric brigand and his devoted men."

At Kaufman, they strolled around town during the afternoon,
and went into camp in the woods nearby. The next morning Bass,
Jackson and Murphy went back to the town, got their horses shod,
Sam and Frank got shaved, went and got their horses after they
were shod, put them in the livery stable and had them fed; and
then went to the hotel and got their dinners. Then they went
over to the east side of town and entered the largest store there
was in town. The object was to find a safe to rob that night. Sam
Bass threw a twenty-dollar bill on the counter and asked the old
man of the store to change it. He took the bill and went to the safe.
When he opened the safe, Sam Bass took a good look into it, and
afterwards said there was scarcely money enough in it to change
the bill. They then returned to camp and started out for Ennis
where they camped a mile from town. Bass and Murphy rode into
Ennis and took a look at the bank. They put their horses in a
livery stable, took dinner at a hotel, and took a second look at
the Ennis bank and concluded that it was fixed too well to rob.

While at Waco, Bass went to the Ranche saloon, and after taking a drink threw a twenty-dollar gold piece on the counter. This
was the last of the money obtained in the Union Pacific robbery, and
he remarked afterwards, "It is all gone, now, and that is all the
good it has done me."

CHAPTER XX

THE LAST FIGHT

Inspecting the Round Rock Bank—Major Jones and the Rangers—
Going to Town After Tobacco and Things—The Fight Begins—
Sheriff Grimes Killed — Sharp Conflict on the Street — Bass
Pierced With a Bullet—Barnes Shot Dead While Mounting His
Horse—Escape of Bass and Jackson—Murphy Appears on the
Scene of Conflict.

As we have seen, according to Murphy's statement, the gang
reached the vicinity of Round Rock Sunday evening, July 14th, and
there went into camp. The next night they moved their camp near-
er new Round Rock, south of the grave-yard, near some negro quarters. Here they remained, resting their horses and visiting the
town, going into the bank and taking a good look at the situation.
Bass and Murphy both had $5 bills changed at the bank. Murphy
delayed the robbers as long as he could, in order to give Major
Jones time to arrive. Finally, when their horses were fully rested and the bank and all its surroundings had been thoroughly
examined, Bass fixed upon the following plan of robbery:

They were all to go to the bank on foot, leaving their horses
hitched in an alley near the bank. Barnes was to give the cashier
a $5 bill to change—the last they had, so it is said—and while he
was doing this Bass was to go behind the counter and level his pistol at the cashier and make him hold up his hands, when Barnes
would jump over the counter, take the money and put it in a sack.
In the meantime Jackson and Murphy were to stand in the door of
the bank to keep anybody from coming in. After getting the money
they were to move out the San Saba road a short distance, then turn
to the right, go up west of Georgetown and make their way up to
Denton, where they proposed to kill Deputy Sheriffs McGing and
Wetzell, of Denton county.

They swore death to Billy Scott, the witness, if they had to ride
to Dallas for hi m. Saturday, July 20th, was the day fixed upon for
the robbery. It was to be in the afternoon just as the bank was to be
closed, at which time they expected all the business men would
have deposited their money.

In the meantime Major Jones had received Murphy's letter from
Belton and Georgetown at Austin. As soon as the letter reached
him he immediately sent to Lieutenant Reynolds, in command of
a squad of rangers at Lampasas, to meet him at Round Rock the

next morning. Three men were also sent to Round Rock early on
the morning of the 18th, and the Major himself followed on the
first train. He took with him Maurice Moore, deputy sheriff of
Travis county, whom he happened to meet· on the street as he was
going to the depot. Moore was formerly a sergeant in his command.
Arriving at Round Rock he went to the postoffice, expecting a letter
from Murphy, but found none. He then warned the banker that the
robbers were in the vicinity, and would probably attempt to rob
them. He called on Deputy Sheriff Grimes, who was once a member
of his command, and took him and Mr. Albert Highsmith into his
confidence. They sent spies out to search the country round for the
robbers' camp. At nightfall, having heard nothing of the robbers,
and not knowing but what they had passed on to Austin, or concluded to strike the train at some other point, Major Jones notified
Captain Hall and the Sheriff and United States Marshal to look out
for them in Austin, and telegraphed the railroad officers at Hearne
and Austin to have the trains guarded.

That night he had his men concealed at the depot to protect the
train, and also had the town thoroughly patrolled. Next morning
his spies were out by daylight, searching the country for the camp.
His men were instructed particularly to keep a lookout about the
bank. About noon, having learned that Lieutenant Reynolds had
removed from Lampasas to San Saba, and fearing that he would
not arrive in time, he telegraphed to Austin for Captain Hall,
who arrived at 2 o'clock p. m. After consultation they telegraphed
to Austin for Lieutenant Armstrong and some of Hall's men, as it
was supposed the robbers numbered seven or eight men.

The critical hour was not at hand. But we turn back for a moment to follow the movements of the robbers as they approached
the scene of deadly conflict.

"Friday morning, the 19th, says Murphy, "Frank Jackson and
me went into town to look for rangers, as Sam Bass said he saw two
rangers who lookd like cow-boys. So we went to see, and we could
not find any, and at eleven o'clock we left town and reported to
Bass. We then smoked awhile, and agreed that all should go to
town after some tobacco and things, as we should rob the bank
next day. When we arrived at the old town I suggested remaining
there to see if I could learn anything of the rangers. They agreed
to this, and Bass, Barnes and Jackson went into the new town."

Murphy's work was now accomplished. What immediately
follows is best told by Deputy Sheriff Maurice Moore: "About
4 p. m.," he says, "I was standing in front of Smith's livery stable,
and three men passed up the street. Smith remarked to me,
'There go three strangers.' I noticed them carefully and thought
one of them had a six-shooter under his coat. The others were

carrying saddle-bags. They looked at me rather hard and went across the street into a store. I walked up the street to where Grimes, the Deputy Sheriff of Williamson county was standing, and remarked to him, 'I think one of those men has a six-shooter on.' Grimes remarked to me, 'Let me go over and see.' We walked across the street and went into the store. Not wishing to let them know I was watching them, I stood up inside the store door with my hands in my pockets, whistling. Grimes approached them carelessly and asked one if he had not a six-shooter. They all three replied, 'Yes,' and at the same instant two of them shot Grimes and one shot me.

After I had fired my first shot I could not see the men on account of the smoke. They continued shooting and so did I, until I fired five shots; as they passed out I saw one man bleeding from the arm and side; I then leaned against the store door, feeling faint and sick, and recovering myself, I started on and fired the remaining shot at one of the men.

"Having lent one of my pistols to another man the day before, I stopped and reloaded my pistol, went into the stable and got my Winchester and started in pursuit of them, and was stopped by Dr. Morris, who said, 'Hold on; don't go any further, for if you get over-heated your wound may kill you;' I stopped and gave my Winchester to another man; went with the Doctor and Judge Schultz to the hotel; Grimes did not have time to draw his pistol; six bullet holes were put through his body." Sheriff Moore was shot through the left lung.

In the meantime the three rangers had come from where they were stationed and fired on the robbers as they retreated across the street. Major Jones, who was coming from the telegraph office when the firing began, ran to the Robinson corner, when seeing the situation, he called on his men, drew his pistol, ran up the street, and when within fifty yards of the robbbers, commenced firing upon them. One of the robbers turned as he reached the corner around which they are retreating and fired deliberately at Major Jones, the ball passing over his head and entering the wall of a building in his rear. At this time the excitement in the town was fearful to witness. Men were running in every direction, some to get out of range of the whistling bullets and take shelter behind a friendly corner, tree or post; others to get such arms as they could lay their hands on and join in the fight; women and children were screaming and flying from the houses between and around which the robbers were retreating. All this presented a scene which beggars description. The robbers retreated across the street, half way up the square and down the alley, at the lower end of which their horses were hitched, closely pursued and constantly fired at by rangers and citizens, but taking shelter and firing back at their pursuers at every convenient place. When half way down the alley, Bass received his second wound, the one which caused his death.

This fatal shot which ended the wild career of the robber chieftian, was fired by George Harrell, a ranger.

Just as the robbers reached their horses, R. C. Ware, one of the rangers, took deliberate aim at one of them and shot him through the head, killing him instantly. As the other two mounted and ran off, Major Jones, Ware, and J. F. Tubbs, a one-armed citizen who had taken Grimes' pistol and joined in the fight, fired several shots at them but without effect. F. L. Jordan fired at the robbers from the back door of his store as they ran down the alley. Albert Highsmith shot at them from the back yard of his stable, and might have killed one of them had not a cartridge shell hung in his Winchester.

Captain Hall was at the hotel lying down when the fight started, but was quickly on the spot with Winchester and pistol in hand, mounted a horse which happened to be near and, accompanied by the three rangers, one of whom rode the dead robber's (Barnes) horse, gave chase to the flying robbers. Several citizens who had horses at hand went with him. As soon as Major Jones could get a horse he, accompanied by Major Dick Mangrum and several other citizens, went in pursuit of the robbers also, but they did not go more than two or three miles before the old plug which the Major had gotten from the livery stable played out and the party returned.

Capt. Hall pursued the robbers until the trail was lost in the brush and then returned to town, as it was too near night and his horses were too nearly broken down to follow further.

When the flying robbers passed through old Round Rock, Jim Murphy was still there and saw them as they dashed by. He says:

"I was sitting in a door at old Round Rock as they came by, and Frank was holding Bass on his horse. Bass looked pale and sickly, and his hand was bleeding, and he seemed to be working cartridges into his pistol. Jackson looked at me as much as to say, Jim, save yourself if you can. Barnes had been killed instantly. I then saw Major Jones go by, and hallooed to him, but he did not hear me. I then went into the new town; there was a good deal of excitement, and some one asked who the dead man was. I said if it is the Bass gang, it must be Seaborn Barnes. Some one asked how they would know. I said he has got four bullet holes in his legs—three in his right and one in his left leg, which he got at Mesquite. They found the wounds, and was going to arrest me, when Major Jones came up, and shortly after recognized me, and I went down with him and identified the dead body as that of Seaborn Barnes."

About two hours after the fight, Lieutenant Reynolds arrived with ten men, having ridden from San Saba, a distance of one hundred miles since seven o'clock the evening before. He left his men a mile or two out of town and came in to report to Major Jones secretly before bringing his men in.

Later in the evening Lieutenant Armstrong's party from Austin arrived.

CHAPTER XXI

CAPTURE AND DEATH

Pursuit of the Robbers — Bass Discovered Under a Live Oak — Fatally Wounded—Taken to Town—Attempt to Secure a Confession—His Dying Statements—Game to the Last—Death-Bed Scene.

It was fully known that one of the robbers who had escaped was badly wounded, as he made two attempts before he was able to mount his saddle. He was also seen holding up his hand as he dashed away, and apparently maintained his seat with great difficulty. As we have already seen, Murphy was very positive that this was Bass.

Major Jones was therefore greatly elated with the prospect of his capture early the next morning. As soon as it was light, Sergeant Nevill of Lieutenant Reynold's company, with eight men was sent out to look for the trail and continue the pursuit. Deputy Sheriff Tucker, of Georgetown, was sent along as guide, as he was thoroughly acquainted with the country. The party proceeded to the point where the trail was lost the evening before. This was about four miles from town. Soon after arriving there, a man was noticed lying under a tree, not far from the new railroad, but as there were some mules grazing near by and as the railroad hands were not far distant Sheriff Tucker said it must be one of the hands and no further attention was paid to him.

The lost trail was found and followed until it divided. After wandering about for some time Sergeant Nevill again emerged upon the prairie, and meeting one of the railroad hands asked him if he had seen a wounded man in the vicinity. He replied that there was a man lying under "that tree out there," pointing to the man seen before, "who was hurt, and who said that he was a cattle man from one of the lower counties, and had been in Round Rock the day before and getting into a little difficulty, had been shot." Sergeant Nevill at once approached the tree and when within about twenty feet of it the wounded man held· up his hand and said: "Don't shoot; I am unarmed and helpless; I am the man you are looking for; I am Sam Bass!"

He had parted with Jackson the evening before after giving him his rifle, pistol and pocket-book, feeling convinced that he would never need them again. During the long, weary hours of the night he lay in the silent woods alone, his body wracked by pain and his mind harrassed with the hopelessness of escape.

In the morning he dragged himself out in the hope of obtaining help. Soon after a negro came by with a team and he tried to hire him to haul him away and secrete him, but failed.

Major Jones was notified, and in company with Dr. Cochran, a physican of Round Rock, went out with an ambulance to bring the prisoner in. After an examination of his wounds, the Doctor pronounced them fatal and assured the bold bandit that his last hour was close at hand. Bass was fully persuaded of the truth of the Doctor's opinion and expressed no hope of recovery.

He was placed in the ambulance and taken to Round Rock, and at once it was telegraphed abroad that he was dying. The fatal bullet had entered the small of the back and come out in front. Much attention was shown him by Major Jones and all present, and nothing was left undone to soothe his pains, in hope of gaining his confidence and softening his fixed determination to reveal nothing against his confederates who were still at large.

He continued in a sinking condition during Saturday, but Sunday morning seemed much better and at once began to entertain a hope of recovery. His physician besought him to make a confession, as he must soon die and appear before the Great Judge. But the wounded robber turned and looking coolly up at the Doctor, said, "don't you be too sure of that."

Major Jones tried every inducement to secure important statements from him, and some one was constantly present with paper and pencil in hand, to write down his utterances, but nothing valuable in the way of evidence, escaped from his lips. His self-control and resolute purpose to remain faithful to his friends were wonderful. Though surrounded by a number of shrewd men and though constantly interrogated by one whose long experience in the capture of outlaws had given him a keen insight into their disposition and made him an adept in handling them, and though the death damp was gathering upon his brow, and final dissolution was wrenching body and spirit apart, yet his wonderful shrewdness and sagacity of

instinct remained intact.   Had he been seated at a camp fire in his
old fastness, surrounded by his pals and sound in health and limb,
he could not more successfully have parried the interrogatives put
to him and thwarted the purpose of his captor.

"I tried every conceivable plan," said Major Jones, "to obtain
some information from him, but to no purpose.   About noon on
Sunday, he began to suffer greatly and sent for me to know if I
could not give him some relief.   I did everything I could for him.
Thinking this an excellent opportunity, I said to him, 'Bass, you
have done much wrong in this world, you now have an opportunity to do some good before you die by giving some information
which will lead to the vindication of that justice which you have
so often defied and the law which you have constantly violated.'   He
replied, 'No, I won't tell.'   'Why won't you?' said I.   'Because it is
agin my profession to blow on my pals.   If a man knows anything
he ought to die with it in him.'   He positively refused to converse
on religion and in reply to some remark made, he said 'I am going to Hell, anyhow.'   I made a particular effort to obtain some information from him in regard to William Collins.   I asked him if he
was ever at Collins' house.   He said no.   I then put the question
in a different form, saying 'where did you first see Will Scott?'
He replied at Bob Murphy's.   I then said, 'You saw him at Green
Hill's too, didn't you?'   He replied, 'yes.'   These answers were
not of any consequence, but I then said, 'when did you see him at
William Collins?'   He said, 'I don't remember, as I never paid attention to dates, being always on the scout, I only saw him these
three times.'   This answer was important, as it fixed the fact that
Bass was at Collins' house.   But this was the only statement of
any importance which he made.   All his other statements were of
facts well known or concerning individuals beyond the reach of future justice."

Among these statements he said:

"I am twenty-seven years old, have brothers John and Denton, at
Mitchell, Indiana.   Have been in the robbing business a long time.
Had done much business of that kind before the U. P. robbery last
fall."

Q.—How came you to commence this kind of life?

A.—Started out sporting on horses.

Q.—Why did you get worse than horse racing?

A.—Because they robbed me of my first $300.

Q.—After they robbed you, what did you do next?

A.—Went to robbing stages in the Black Hills—robbed seven.
Got very little money.   Jack Davis, Nixon and myself were all that
were in the Black Hills stage robberies.   Joel Collins, Bill Heffrige,

Tom Nixon, Jack Davis, Jim Berry and me were in the Union Pacfic robbery. Tom Nixon is in Canada. Have not seen him since that robbery. Jack Davis was in New Orleans from the time of the Union Pacfic robbery till he went to Denton to get me to go in with him and buy a ship. This was the last of April, 1878. Gardner, living in Atascosa county, is my friend. Was at his house last fall. Went to Kansas with him once. Will not tell who was in the Eagle Ford robbery besides myself and Barnes. When we were in the store at Round Rock, Grimes asked me if I had a pistol, I said yes; then all three of us drew our pistols and shot him. If I killed Grimes it was the first man I ever killed. Henry Collins was with me in the Salt Creek fight four or five weeks ago. Arkansas Johnson was killed in that fight. Don't know whether Underwood was wounded or not at Salt Creek fight. 'Sebe' Barnes, Frank Jackson and Charles Carter were there. We were all set afoot in that fight, but stole horses enough to remount ourselves in three hours, or as soon as dark came; after which we went back to Denton. Stayed there until we came to Round Rock.

Q.—Where is Jackson now?
A.—I don't know.
Q.—How do you usually meet after being scattered?
A.—Generally told by friends.
Q.—Who are these friends?
A.—I will not tell.

This was his usual reply to questions which he did not wish to answer, and was in the most deliberate manner possible.

Even in the midst of his intense agony on Sunday afternoon he clung to the delusion that he would recover. But about twenty minutes before his death, when warned by his physician that dissolution was near at hand, he calmly replied, 'let me go.'

A few minutes later he said to his nurse, "the world is bobbing around me." His pains had ceased and he rested at ease. There were a few gasps and he was dead. This was at 4 p. m., Sunday, June 21st.

The next day the body was interred at Round Rock. And thus the earth gathered back to her bosom one who had lived to harass and torment his kind.

What is known in regard to the rest of the band is easily told. Jackson made good his escape, reaching Denton county two hundred miles distant after a three days ride. But his capture may occur at any time. Underwood had not been seen since the Salt Creek fight, and his whereabouts are unknown. Carter. who joined Bass towards the last, is said to have been sent out of the country

by his father. The two Collins, under indicment by the grand
jury, are still in concealment. Jim Murphy received his reward,
and is now at his home near Rosston in Denton county.

---◇---

CHAPTER XXII

NOTES AND COMMENTS

Excellent Work Accomplished By The Authorities—A Word Of
Justification for Captain Peak — Local Authorities — Force of
Detectives Needed — Prevention of Crime — Evil Influences of
Horse-Racing and Gaming — Need of Education — No Profit in
Crime.

It is now evident that there was some remissness in not making
a more prompt and determined effort to hunt down the train
robbers before they had so successfully repeated their outrages.
It must also be admitted that the great campaign against Bass in
Denton and Wise counties was not a success. But after all the
State of Texas has reason to congratulate herself on what has been
accomplished. The first robbery was committed February 22, and
before that date in July the leader of the gang and two of his leading
accomplices had been laid in bloody graves, three others had been
convicted and sent to the penitentiary, one other is still in jail, five
others have been indicted and arrested as accessories, and are now
out on bond, two others under arrest were allowed their liberty for
services rendered the State. Only two principal members of the
band, Jackson and Underwood have made their escape.

This makes an excellent showing for our authorities, and speaks
well for the determined and efficient manner in which the band has
been hunted down amid wild woods and a sparsely populated country. It was thought that much was accomplished by the pursuers
of the Union Pacific robbers and yet only three out of six of the
robbers were captured. But in the case of the Texas train robbers,
six out of eight of the principals in the crimes have ben killed or
covicted. If we number Green Hill and William Collins among the
principals, the former is safe in jail at Austin, and the appearance
of the latter before court is secured by a $15,000 bond.

In regard to the failure of the campaign against Bass, justice
requires that Captain Peak should be set right before public opinion.
For all that occurred between April 24th and May 21st he is responsible to the just expectations of the public. But it must be remem-

bered that on May 21st the contract was entered into with James
Murphy to betray Bass, and in furtherance of this purpose it was
arranged that Captain Peak should hold his force at or near Dallas.

The agreement with Murphy being necessarily concealed from
the public some unjust criticism was indulged in against Captain
Peak for his inaction. But now that the whole plan has been un-
covered, it is plain that he was fully justified in holding his command stationary at a convenient point where Murphy could read-
ily reach him with his communications. But on account of the
pressure of public opinion he was compelled to continue more actively in the pursuit of the gang than was deemed desirable. This
led to the pursuit of the band into Wise county and the fight at
Salt Creek the only fight in the long pursuit which was attended
by a good result.

In regard to the action of Sheriffs and local authorities there
is also need of a word of explanation and justification. The laws
of the State make no provision for the expenses of a sheriff's
posse engaged in a prolonged pursuit of outlaws. It matters not
how far or how long a sheriff may ride, or how many armed and
mounted men he may employ to assist him, or how much money he
may spend on the trip, he is only allowed one dollar each for the
criminals he may capture, and that after conviction. The Sheriff
of Dallas county declared that he could not pursue that gang who
robbbed the trains in the county, because he could not afford it.

It will be a very important question for our next Legislature
to consider, whether the laws should not be amended in this respect, and our local authorities strengthened and made far more
efficient by a proper provision for necessary expenses in cases of
emergency.

Another very important question which should also be considered, is whether it would not be well for the State to employ a
regular force of detectives to ferret out and secure the arrest of
criminals.

It is well known that there are many outlaws and fugitives
from justice in the State. A list of four thousand was published
not long since, and this did not include reports from a large number of counties. Many criminals have also escaped from other
States and fled to Texas for refuge. Against these outlaws the
State police is performing very efficient service. But their efforts
should be supplemented by detectives, working secretly among the
outlaws, discovering them in their hiding places and securing the
proper chain of evidence for their conviction.

There has been too much of a tendency heretofore to rely upon
revolvers and bold riders. But it should be remembered that the

pell mell drive after Bass accomplished little, while to a spy and to a betrayer we are primarily indebted for the capture of leading members and final overthrow of the band. The method pursued for the breaking up of the Molly Maguires is very instructive in this connection. This was, perhaps, the worst combination of outlaws ever known in this country. When Franklin B. Gowan, president of the Reading railroad, determined to break it up, he employed the ablest of Pinkerton's detectives to accomplish the task. They went into the counties infested by the members of the organization, and continued their efforts until they arrested and convicted more than sixty of the outlaws, many of whom were hung. Similar service against the thousands of criminals who infest the State, would undoubtedly be attended with most important results.

But back of all considerations of the best method of pursuing criminals, lies a still more important question not only for the State Government but for society, and that is how to prevent men from becoming criminals. The occasional case of Sam Bass' criminal career is easily stated. He, himself, and his employes, and neighbors say, that it was the purchase of the race mare. Horse-racing soon lead him into a career of idleness and dissipation, and from that the descent to open outlawry was easy.

That the influence of horse-racing and gaming was ruinous in this instance, is a plain fact, testified to in the dying confession of a slain outlaw. That it is almost invariably demoralizing must be admitted by all. It should not, therefore, be encouraged by the laws of the State or voluntary organization of the people. When our fair associations devote the larger part of their premiums to horse-racing, and when they admmit all forms of light gaming to a place among their exhibitions, they do more to demoralize the young and to impair the moral integrity of a community than they do to promote its industrial and agricultural interests.

Again, it should not be forgotten that Sam Bass was uneducated. For this Texas was not responsible, for he was a young ignoramus thrust upon us by Indiana. But ignorance is a fruitful source of crime and costs the State infinitely more than education. We can never prevent crime until we go back to the sources of intellectual and moral life. The work must be begun near the cradle, and pursued with never wearying vigilance until the character is fully matured and the mind thoroughly imbued with the highest and noblest principles.

In conclusion, one word to the young. The history of these robbers is an appalling argument against such a life. Their career was very short.. They were driven from the face of their fellow-

men. Their ill-gotten gains did them no good. Vengeance came
swift and terrible, and in a few days, or at most a few short
months, they were in° bloody graves, or imprisoned at hard labor
and forever disgraced. Mankind rejoiced at their fall. For those
who lift their hands against law and order, the world has only
condemnation disgrace and death.

To mark the grave of that restless man a simple monument stands
in the little town of Round Rock bearing the inscription

Samuel Bass
Born July 21st, 1851
Died July 21st, 1878

A Brave Man Reposes in Death Here. Why Was He Not True?

Sam Bass was true to his friends and his convictions, but what of
Jim Murphy? That man, hated of all men, despised even by the
rangers whom he had served, returned to Denton. Words cannot
express the supreme contempt and hatred for the man (used for
classification only) who, like a rattlesnake, turned and bit the one
who befriended him. A guilty conscience weighing heavily upon
him caused him to seek protection from the sheriff when his dis-
torted imagination led him to believe Frank Jackson was lying
around in the Elm Bottoms waiting for a chance to kill him. The
sheriff granted him permission to take up his abode in the jail.
However, his stay in his jail home was of short duration, as in a
few weeks his ignominious career was brought to a close by suicide.